Practical Python Programming for Data Scientists

Practical Python Programming for Data Scientists

A. Suresh, N.Malarvizhi, Pethuru Raj,
and E. A. Neeba

ARCLER
PRESS

www.arclerpress.com

Practical Python Programming for Data Scientists

A. Suresh, N.Malarvizhi, Pethuru Raj, and E. A. Neeba

Arcler Press
224 Shoreacres Road
Burlington, ON L7L 2H2
Canada
www.arclerpress.com
Email: orders@arclereducation.com

ISBN: 978-1-77469-158-8 (Hardcover)

ABOUT THE AUTHORS

A. Suresh, Ph.D. works as the Associate Professor, Department of the Computer Science and Engineering in SRM Institute of Science & Technology, Kattankulathur, Chengalpattu Dist., Tamil Nadu, India. He has been nearly two decades of experience in teaching and his areas of specializations are Data Mining, Artificial Intelligence, Image Processing, Multimedia and System Software. He has published three patents and 90 papers in International journals. He has book authored "Industrial IoT Application Architectures and use cases" published in CRC press and edited book entitled "Deep Neural Networks for Multimodal Imaging and Biomedical Application" published in IGI Global. He has currently editing three books namely "Deep learning and Edge Computing solutions for High Performance Computing" in EAI/Springer Innovations in Communications and Computing, "Sensor Data Management and Analysis: The Role of Deep Learning" and "Bioinformatics and Medical Applications: Big Data using Deep Learning Algorithms" in Scrivener-Wiley publisher. He has published 15 chapters in the book title An Intelligent Grid Network Based on Cloud Computing Infrastructures in IGI Global Publisher and Internet of Things for Industry 4.0 in EAI/Springer Innovations in Communication and Computing. He has published more than 40 papers in National and International Conferences. He has served as editor / reviewer for Springer, Elsevier, Wiley, IGI Global, IoS Press, Inderscience journals etc... He is a member of IEEE (Senior Member), ISTE, MCSI, IACSIT, IAENG, MCSTA and Global Member of Internet Society (ISOC). He has organized several National Workshop, Conferences and Technical Events. He is regularly invited to deliver lectures in various programmes for imparting skills in research methodology to students and research scholars. He has published four books in Indian publishers, in the name of Hospital Management, Data Structures & Algorithms, Computer Programming, Problem Solving and Python Programming and Programming in "C". He has hosted two special sessions for IEEE sponsored conference in Osaka, Japan and Thailand.

N. Malarvizhi, PhD currently working as the Professor in the Department of Computer Science and Engineering at Vel Tech Rangarajan Dr. Sagunthala R&D Institute of Science and Technology, Chennai-62, Tamilnadu, India. She is having more than 18 years of teaching experience. She has written a book titled "Computer Architecture and Organization", Eswar Press, The Science and Technology Book Publisher, Chennai. She serves as a reviewer for many reputed journals. She has published numerous papers in International Conferences and Journals. Her area of interest includes Parallel and Distributed Computing, Grid Computing, Cloud Computing, Big Data Analytics, Internet of Things, Computer Architecture and Operating Systems. She is a life member of Computer Society of India (CSI), Indian Society for Technical Education (ISTE), IARCS and IAENG. She is a Senior Member of IEEE and IEEE Women in Engineering (WIE). She is a Member of Association for Computing Machinery (ACM) and The Institution of Engineering and Technology (IET).

Pethuru Raj has been working as the chief architect in the Site Reliability Engineering (SRE) Center of Excellence, Reliance Jio Platforms., Bangalore. He previously worked as a cloud infrastructure architect in the IBM Global Cloud Center of Excellence (CoE), IBM India Bangalore for four years. Prior to that, He had a long stint as TOGAF-certified enterprise architecture (EA) consultant in Wipro Consulting Services (WCS) Division. He also worked as a lead architect in the corporate research (CR) division of Robert Bosch, Bangalore. In total, He have gained more than 17 years of IT industry experience and 8 years of research experience. He obtained his PhD through CSIR-sponsored PhD degree in Anna University, Chennai and continued the UGC-sponsored postdoctoral research in the department of Computer Science and Automation, Indian Institute of Science, Bangalore. Thereafter, He was granted a couple of international research fellowships (JSPS and JST) to work as a research scientist for 3.5 years in two leading Japanese universities. Regarding the publications, He have published more than 30 research papers in peer-reviewed journals such as IEEE, ACM, Springer-Verlag, Inderscience, etc. He has authored 7 books thus far and He focus on some of the emerging technologies such as IoT, Cognitive Analytics, Blockchain, Digital Twin, Docker-enabled Containerization, Data Science, Microservices Architecture, etc. He has contributed 25 book chapters thus far for various technology books edited by highly acclaimed and accomplished professors and professionals. The CRC Press, USA had also released his first book titled as "Cloud Enterprise Architecture" in the year 2012 and you can find the book details in the page http://www.crcpress.com/product/isbn/9781466502321 He has edited and authored a book on the title" Cloud Infrastructures for Big Data Analytics" published by IGI International USA in March 2014. A new book on the title" Smarter Cities: the Enabling Technologies and Tools" by CRC Press, USA, is to hit the market in the month of June 2015. He has collaborating with a few authors to publish a book on the title" High-Performance Big Data Analytics" to be published by Springer-Verlag in the year 2015.

E. A. Neeba, currently working as an Assistant Professor in the Department of Information Technology at Rajagiri School of Engineering & Technology, Kochi, Kerala, which is affiliated to the A.P.J Abdul Kalam Technological University, Kerala. She received her doctoral degree from Vel Tech Rangarajan Dr. Sagunthala R&D Institute of Science and Technology, Chennai, Tamil Nadu. She completed her Masters in Computer Science & Engineering from SRM Institute of Science and Technology, Chennai. Her research interests include Analysis of data, Data Mining and Big Data, knowledge representation, and ontology, both from the theoretical perspective and their application to natural language understanding, reasoning, information visualization, and interoperability. Having a rich industrial experience of around 10 years prior to joining academia, and also, she has publications in around 10 SCI/ SCIE/Scopus indexed international journals and a few national journals. An active participant in various conferences and workshops on data mining, she is currently involved in several projects in this field. She was entrusted with leadership positions such as the Accreditation coordinator for the college, and Head of the Quality Cell, besides organizing various national and international events.

TABLE OF CONTENTS

LIST OF FIGURES

LIST OF TABLES

LIST OF ABBREVIATIONS

AI	Artificial Intelligence
ANNs	Artificial Neural Networks
ARIMA	Autoregressive Integrated Moving Average
BI	Business Intelligence
CMS	Content Management Systems
CPS	Cyber-Physical Systems
CV	Computer Vision
CWI	Centrum Wiskunde and Informatica
DApps	Decentralized Applications
DBMS	Database Management Systems
DBSCAN	Density-Based Spatial Clustering of Applications with Noise
DL	Deep Learning
DNNs	Deep Neural Networks
DS	Data Science
DSS	Decision Support Systems
EDA	Event-Driven Architecture
ELT	Extract, Load, and Transform
EM	*Expectation-Maximization*
GAN	Generative Adversarial Network
GUIs	Graphical User Interfaces
IoT	Internet of Things
K-NN	K Nearest Neighbors
LSTM	Long Short Term Memory
ML	Machine Learning
MSA	Microservices Architecture
NLP	Natural Language Processing
NN	Neural Network

PCA	Principal Component Analysis
PEMDAS	Parentheses, Exponentiation, Multiplication, Division, Addition, Subtraction
Q&A	Question and Answering
RF	Random Forest
SARIMA	Seasonal Autoregressive Integrated Moving Average
SMEs	Subject Matter Experts
SNA	Social Network Analysis
SVMs	Support Vector Machines
VAR	Vector Autoregression

PREFACE

Data science (DS) is a fast-emerging field of study and research. It leverages integrated data analytics (big, fast, and streaming analytics) platforms and artificial intelligence (AI) (machine and deep learning (ML/DL), computer vision (CV), and natural language processing (NLP)) algorithms extensively to extract actionable insights out of burgeoning data volumes in time. There are several things happening concurrently in the IT domain.

1. With the surging popularity of digitization and edge technologies, there is a huge surge in digitized entities/smart objects/sentient materials in and around us.
2. The device ecosystem is growing rapidly with the ready availability of purpose-agnostic and specific devices for personal, professional, and social purposes.
3. There is a faster maturity and stability of scores of connectivity technologies, it is anticipated that there will be billions of connected embedded systems
4. With the wider acceptance and adoption of the microservices architecture (MSA) and event-driven architecture (EDA) patterns for producing and sustaining enterprise-scale applications, there is a rapid rise in the number of usable and reusable event-driven microservices
5. With the purposeful interactions between digitized entities, connected devices and interoperable microservices, there is a massive amount of multi-structured data getting produced, collected, cleansed, and crunched meticulously
6. There are data analytics and science platforms in plenty to discover and disseminate knowledge out of data heaps

The implications of the various digital technologies and tools are given below. In the years ahead, we will be bombarded with

- Millions of microservices;
- Billions of connected devices;
- Trillions of digital entities.

Thus, the smart leverage of pioneering digitization and digitalization technologies is to result in unprecedented growth in data production. However, if generated data are not subjected to a series of deeper investigations to squeeze out actionable insights in time, then the data goes off wastefully. There is a realization that data is the new oil intrinsically capable of fueling the world for long. Precisely speaking, every data (internal and external) being produced by any establishment has to be meticulously collected, processed, and mined in order to realize the much-needed cognition not only for human beings in their everyday decisions, deals, and deeds but also for devices and machines to be intelligent in their operations, outputs, and offerings.

Data science plays a very vital role in shaping up the process of transitioning data into information and into knowledge. As business enterprises, organizations, governments, IT companies, and service providers are keenly becoming data-driven, the role and responsibility of data scientists are bound to go up significantly. There are several enabling frameworks, libraries, tools, accelerators, engines, platforms, cloud, and edge IT infrastructures, optimized processes, patterns, and best practices to simplify and streamline data science tasks for data scientists.

Python is emerging as the leading programming language for data science projects. Python brings in a number of technical advantages for the successful implementation of data science applications. Due to the ready availability of several libraries for facilitating the development of data science services, Python is turning out the programming language of choice for data science. The following libraries are enabling data science applications and are made available in Python:

1. **NumPy:** This is a library that makes a variety of mathematical and statistical operations easier and faster. This is also the basis for many features of the Pandas library.
2. **Pandas:** This Python library is created specifically to facilitate working with data. This is one of the gamechangers for the tremendous success of data science projects.
3. **Matplotlib:** This is a visualization library that makes it quick and easy to generate charts from data.
4. **Scikit-Learn:** This is the most popular library for machine learning (ML) work in Python.

The book starts with a couple of chapters on data science and machine learning (ML) topics. Thereafter, the chapters are focusing on the fundamental and foundational aspects of Python programming language. All kinds of language constructs are accentuated and articulated for the benefit of programmers with all the practical details. There are dedicated chapters for producing machine learning applications. The gist of the book is to clearly explain how Python simplifies and speeds up the realization of next-generation data science applications. All the specific libraries towards data science are given the required thrust in order to empower our esteemed readers with all the right and relevant information. This book is being prepared with the intention of empowering data scientists with all the vital details about programming using the Python language.

—Pethuru Raj, PhD

CHAPTER 1

THE DISTINCTIONS OF PYTHON LANGUAGE

CONTENTS

The inventor of Python says that the joy of coding using Python should be in seeing concise, precise, and readable classes that express a lot of action in a small amount of clear and lean code. Python is a programming language that lets you work quickly and integrate systems more effectively.

1.1. INTRODUCTION

With digitization and digital technologies and tools becoming matured and stabilized fast, the much-expected real business transformation is all set to become a grandiose reality across industry verticals soon. Digital innovations and disruptions are frequently and feverishly happening in the business domain these days. Not only businesses but also common people are also increasingly experiencing digitally empowered living. With digital data getting accumulated and stocked in cloud storages, the challenge is how data scientists are able to crunch digital data to extract actionable insights in time. The widespread acceptance and adoption of digital paradigms such as the establishment of software-defined cloud environments, artificial intelligence (AI) algorithms, digital twins, blockchain technology, the internet of things (IoT), 5G communication, microservices, and event-driven architectures (EDAs), etc., come handy in making sense out of digital data. The knowledge discovered gets disseminated to concerned systems and decision-makers in time so that appropriate counter measures can be considered and taken with all the confidence and clarity.

There are several noteworthy technologies-inspired transformations. Businesses, government organizations, institutions, establishments, and cities are constantly digitally empowered to bring forth premium and pioneering services to their constituents. As we are tending towards the digital world, we hear, read, and even experience a bevy of digital applications. Other buzzwords in the digital era include digital intelligence and economy. We need a path-breaking programming language to build flexible, fabulous, and futuristic software packages. Python is being recognized as the top programming language for constructing digitally transforming applications. In this chapter, we would like to throw some light on this innovation, which is penetrative, pervasive, and persuasive too.

According to Stackoverflow (https://insights.stackoverflow.com/survey/2019), Python is the most preferred language. That means the majority of developers across the globe use python.

Python has significantly evolved since its creation in 1991 by Guido Van Rossum. In short, Python is an interpreted, dynamic, and high-level programming language that facilitates building a plethora of apps. All kinds of enterprise-grade, service-oriented, cloud-hosted, and event-driven web, machine learning (ML), mobile, embedded, and microservices-centric applications are increasingly developed by Python programming language. In this chapter, we are discussing the prominent domains, which are gaining immensely through the unique capabilities of Python.

1.2. WEB APPLICATION DEVELOPMENT

Python is one of the popular programming languages for web development because Python is being stuffed with a number of frameworks and content management systems (CMS) that significantly simplify web developers' life. Python offers several other benefits, such as simplicity, security, and scalability. Python comes out with the intrinsic support for different data representation and exchange formats. Also, myriads of data transmission protocols are also supported innately. Python gains prominence for its hundreds of libraries to simplify and speed up software programming.

1.3. GAME DEVELOPMENT

Besides web development, game development is also benefiting through the various inherent facilities offered by Python. There are multiple game-specific libraries and automated toolkits to streamline and accelerate the realization of gaming applications.

1.4. ARTIFICIAL INTELLIGENCE (AI) APPLICATIONS

The domain of data science (DS) is gaining surging popularity these days as it has the wherewithal to discover knowledge out of massive amount of multi-structured data. AI algorithms, especially machine and deep learning (ML/DL) algorithms contribute immensely for the spectacular success of the DS domain. Today Python is the preferred language for writing AI applications. There are several high-performance libraries to assist AI developers to quickly come out with competent AI applications. Python's stability and security are being presented as the key differentiators for data scientists to choose Python to code AI and statistical applications. AI, being complicated yet sophisticated applications, can demand high computing power. There are other complications in realizing next-generation AI apps.

Having understood this need, there came a number of unique libraries as enumerated below:

- SciPy for scientific and technical computing;
- Pandas for data analysis and manipulation;
- Keras for artificial neural networks (ANNs);
- TensorFlow for ML tasks and deep neural networks (DNNs);
- NumPy for complex mathematical functions and computation;
- Scikit-learn for working with various ML models.

1.5. GRAPHICAL USER INTERFACES (GUIS)

GUIs are increasingly used by both technical as well as non-technical professionals to interact fruitfully with local as well as remote applications and services. GUI programming has been an important factor for modern applications. Python's comprehensible syntax and the modular programming approach enable creating and sustaining intuitive, informative, and inspiring GUIs.

1.6. COMPUTER VISION (CV) APPLICATIONS

With the widespread utilization of AI algorithms, especially deep learning (DL) algorithms, computers in conjunction with cameras are able to do face recognition, object detection, asset tracking, etc. Not only computers but also industrial machineries and consumer electronics are able to excel in collecting vision data and running vision-based applications. Python has come out with several enabling libraries for simplifying the capture of still and dynamic images and processing them in order to facilitate decision-making and subsequent actions based on vision data.

Natural language processing (NLP) applications: Speech recognition and synthesis are becoming common these days with the leverage of path-breaking AI capabilities. Text processing, which involves a tremendous amount of text, is also gaining importance these days. Text mining is another buzzword. Python's text processing capabilities are simply phenomenal.

1.7. AUDIO AND VIDEO APPLICATIONS

With Python, it is quite easy to handle audio and video files. Tasks such as signal processing, audio manipulation, and recognition can be performed by leveraging specific libraries. For video analytics, Python provides a number of easy-to-use libraries.

1.8. KNOWLEDGE VISUALIZATION APPLICATIONS

Data is the modern asset and fuel to energize and bolster the digital era. Data gets converted into information and into knowledge through a host of digital technologies and tools. DS, which is a collection of digital technologies and tools, predominantly deals with the transition of data into actionable insights. Knowledge thus discovered gets disseminated to execution systems quickly to proceed with the best course of action. Knowledge has to be instantaneously articulated and accentuated to concerned people through knowledge visualization applications in the form of charts, maps, graphs, etc. Python is also primed to produce such applications.

1.9. SCIENTIFIC AND NUMERIC APPLICATIONS

Python is definitely used extensively for creating scientific and technical computing applications. There are powerful libraries exclusively and elegantly catering to this need. There are scientific and numerical tools and libraries such as Pandas, SciPy, and Numeric Python, etc.

1.10. IOT AND CPS APPLICATIONS

All common, cheap, and casual objects in our midst are getting meticulously transitioned into digitized entities through the smart leverage of digitization and edge technologies. That is, all kinds of physical, mechanical, and electrical systems in our personal, social, and professional environments are systematically empowered to be digitized, and this distinct technology-enablement makes them join in the mainstream computing. Similarly, all kinds of consumer electronics, medical instruments, appliances, equipment, kitchen appliances, manufacturing machineries, robots, drones, game consoles, etc., are being connected. That is, all kinds of devices are instrumented and interconnected. There are ways and means for empowering connected devices to be intelligent in their operations. Digitized entities and connected devices, which are going to be in several billions in the years to come, are also integrated with cloud-hosted applications, services, and databases. That is, our daily environments are stuffed by a dazzling array of networked and embedded devices, which are not only integrated locally but also substantially empowered through integration with remote software packages and data sources. Cyber-physical systems (CPS) are conceptually similar to the IoT paradigm. That is, all kinds of ground-level physical systems get hooked to cyberinfrastructure, applications, services, and databases in order to gain extra capabilities to be right and relevant for human beings in their everyday functions. CPSs will be empowered further through a seamless and spontaneous integration with digital twins.

The IoT represents the future Internet comprising all kinds of digitized artifacts in addition to connected electronics and traditional IT systems (servers, storage appliances, desktops, laptops, etc.). The IoT is constantly growing and will be encompassing digitized mechanical, electrical, and physical systems. That is, there will be trillions of digitized entities, billions of connected devices and millions of microservices in the years to come. Such an all-powerful, connected, and complicated Internet poses a series of challenges and concerns. All these improvements and improvisations are leading to the visualization and realization of next-generation IoT applications and services. Such sophisticated and complicated applications can be realized through the intrinsic power of Python.

1.11. DATA ANALYTICS

This is one of the important jobs in emitting out useful insights out of data mountains. Businesses are investing heavily to have data storage and processing infrastructure through private or public or both clouds. There are integrated big, fast, and streaming data analytics platforms. Apache Hadoop, Spark, Samza, Flink, Storm, etc., are the well-known open-source platforms to do batch and real-time data processing. That is, big, fast, and streaming data analytics activities are quickened through the leverage of competent platforms and cloud infrastructures to squeeze out usable insights. Python is one of the popular languages to accomplish data analytics in a simplified and smart manner. PySpark is a great extension/framework to create scalable and data-intensive analyzes and pipelines by utilizing the power of Spark in the background.

1.12. PYTHON FOR BLOCKCHAIN APPS

A blockchain is a "chain of blocks." This chain is presented as a decentralized and distributed database. This does not have a centralized entity to closely monitor and manage the database. Because it is distributed, the aspects such as high availability, lower latency, etc., are easily fulfilled. Each block contains a "hash" and a "index" and information about the particular transaction that took place. All the Blocks in the chain are linked to each other with the "hash" variable. A "hash" contains information of the previous block in the chain, and that's what keeps the entire chain-linked and connected as pictorially represented below.

If any value in any of the blocks in "tampered with," every block thereafter in the chain gets affected and hence stealing information from any of the blocks in an unauthorized manner is a difficult affair. Because any manipulation will cause the hash to change as well and it won't match the hash in the block after it. This will alert the network about the "tampering" and will render the entire chain useless. So for any hacker to successfully hack into a chain, one has to not only change values of one single block but all the blocks before and after it. This wholesome change is nearly impossible. For more details for an end-to-end blockchain example explanation, please visit this page (https://medium.com/swlh/introduction-to-blockchain-with-implementation-in-python-c12f8478a3c4).

Python is turning out to be an excellent programming language for developing blockchain applications, especially decentralized applications (DApps) and smart contracts. There are Python tools and libraries to speed up blockchain apps development. There are several tutorials made available in the Internet servers for easing up blockchain app development using Python.

Thus, all modern applications development is being facilitated by the unique power of the python language. The increasing complexity of microservices-centric, event-driven, and cloud-native applications are being lessened through the smart leverage of Python.

1.13. CONCLUSION

Python is a robust, resilient, and versatile programming language. A growing array of software applications across multiple industry verticals are being coded through Python. A fast-growing list of libraries has made it possible for developers to easily use Python to come out with state-of-the-art applications leveraging cutting-edge digital technologies. More details can be found at the Python home page, which is made available at this URL (https://www.python.org/).

CHAPTER 2

DEMYSTIFYING THE DATA SCIENCE PARADIGM

CONTENTS

2.1. INTRODUCTION

There are several beautiful things happening simultaneously in the IT field. There is an exponential growth of multi-structured data due to the consistent eruption of different and distributed data sources. The fast-growing device ecosystem, the explosion of digitized assets, the emergence of simple websites (web 1.0), social websites (web 2.0), and semantic websites (web 3.0), the journey towards the Industry 4.0 vision, and the faster proliferation of microservices for implementing enterprise-scale software applications are incredibly laying down a stimulating foundation for the data-driven world. The continuous optimization of data-to-information-to-knowledge process with the releases of data analytics platforms and AI toolkits is raising the interest level is being supported across.

And the realization of the truth that data-driven insights and insights-driven decisions and deeds are very critical not only for businesses but also for the total human society is persuasive and pervasive. Software-defined cloud infrastructures are being established and sustained by cloud service provides and enterprise IT teams across the world. Thus, we are being bombarded with integrated platforms, lean processes, high-value products, enabling patterns, facilitating frameworks, knowledge guides, and best practices in addition to highly optimized and organized cloud IT infrastructures. All these spectacular accomplishments clearly tell that the domains of data engineering, analytics, storage, processing, management, and mining are ceaselessly flourishing. We are seeing the grandiose realization of self-learning systems in plenty for solving diverse requirements across industry verticals. In short, the field of data science (DS), the next and intelligent version of data analytics and mining, is progressing fast with the aim of extracting actionable insights out of data volumes in a simpler and quicker manner. In this chapter, we are to discuss about DS and how it is promising to impact the world in a hitherto unheard fashion.

DS can be defined as obtaining decision-enabling, predicting, and value-adding insights out of data heaps. DS, when applied to different fields, can lead to incredible new insights. Precisely speaking, DS transforms the accumulated digital data into data-driven knowledge. Digital data is increasingly not interpreted by an individual anymore. Instead DS relies on machines to interpret, process, and alter it.

2.2. BRIEFING DATA ANALYSIS

Businesses across the world are generating a lot of poly-structured data, which originates from multiple and diverse sources. We have technology-inspired, service-oriented, knowledge-filled, enterprise-scale, event-driven, and cloud-hosted applications (operational, conversational, commercial, analytical, transactional, etc.), and a growing variety of user devices (mobile, wearable, portable, fixed, wireless, handhelds, implantable, etc.). These applications and devices are generating a lot of data. The beauty is that there is something valuable and unique hidden in the data. The challenge is how to squeeze out the right actionable insights out of data heaps in time. There is a need to extract tactical and strategic information and knowledge from data mountains so that business executives and execution systems together steer businesses in the right direction. Data analytics is typically exploratory and being presented as the one capable of answering business-related questions in detail with all the clarity and confidence.

Data analysis, the first subcategory of DS, is all about asking specific questions with the intention of bringing forth usable responses. As the data volume is humungous, finding answers requires enabling tools such as SQL or Python or R. They allow data analysts to aggregate and manipulate data in order to arrive at correct conclusions and to make the right decisions. Data analysts have to produce the right questions using SQL for slicing and searching SQL databases for bringing forth useful answers. SQL works well with structured data. Thus, there are a lot of manual activities involved in collecting, cleansing, and crunching data. Of course, there are data virtualization, ingestion, pre-processing, storage, analytics, and visualization tools for transitioning data into information and into knowledge. DS is the next version of data analytics with a host of things getting automated through powerful tools.

2.3. ENTERING INTO DATA SCIENCE (DS)

As accentuated elsewhere, bringing out useful patterns out of data volumes is the central aspect of DS. Data analytics has been the prime domain blessed with so many frameworks, platforms, tools, accelerators, techniques, and tips to simplify and streamline knowledge discovery and dissemination. DS represents the next-generation data analytics.

As we understand, artificial intelligence (AI) wants some kind of human interaction and is intended to be somewhat human or "intelligent" in the way it carries out those interactions. Thus, it is important to have an intelligent and natural interactions between machines and men. That intelligence is being derived through DS. DS is all about producing insights that enable building intelligent systems. In other words, DS places less emphasis on human interaction but pitches for producing and providing human-like intelligence, strategically sound and sharp recommendations, decisions, and conclusions.

DS is turning out to be an indispensable ingredient for any enterprising business today. With the accumulation of data and the ready availability of algorithms, platforms, and infrastructures, corporates have started setting up and sustaining DS competencies to grow their business and increase customer satisfaction. DS is defined as the futuristic domain of deep study and research to sensibly and scholarly handle a tremendous amount of data. That is, how to make sense and money out of data is the biggest challenge for worldwide data scientists. There are facilitating tools and accelerators, specialized engines and appliances to deal effectively and inspiringly with fast-growing data. It is all about pinpointing unseen patterns for deriving meaningful information, which will go a long way in arriving at future-proof business decisions. DS uses complex machine and deep learning (DL) algorithms to build predictive and prescriptive models. The data used for deeper and decisive analysis can be from multiple sources. There are several data representation and exchange formats. Further on, there are a variety of data transmission protocols. Thus, it is definitely a complicated environment due to multiplicity and heterogeneity. And to bring a kind of homogeneity, pioneering technologies and tools are being leveraged.

Precisely speaking, DS is a blend of various analytics platforms, enabling tools, statistical methods, and AI algorithms with the goal to discover hidden patterns from the raw data.

2.3.1. Business Intelligence (BI) vs. Data Science (DS)

Businesses have been concentrating on business intelligence (BI) for long. With the ready availability of matured BI tools and data visualization dashboards, it is possible to explain what is going on by processing data. That is, it is a kind of exploratory analysis to discover viable insights out of data.

On the other hand, data scientists not only doing exploratory analysis but also apply various ML algorithms to pinpoint the occurrence of a specific event in the future. A data scientist will look at the data from multiple perspectives.

So, DS is predominantly used to make decisions and predictions making use of predictive causal analytics and prescriptive analytics (predictive plus decision science):

1. Predictive Causal Analytics: If you want a model that can neatly predict the occurrence of a particular event in the future, you need to apply predictive causal analytics. For example, if you are providing money on credit, then the probability of customers making credit payments on time in the future is definitely worrisome for you. The way forward is that you have to build a model that can perform predictive analytics on the payment history of the customer to predict if the future payments will be on time or not.
2. ***Prescriptive Analytics***: If you want a model that has the inherent intelligence of taking its own decisions and the ability to modify the model with dynamic parameters, you certainly need prescriptive analytics capability for it. That is, prescriptive analytics not only predicts but prescribe a set of actions and associated outcomes to achieve the predicted. In the case of self-driving cars, the data gathered by cars can be used to train self-driving cars. This training emits out insights. Cars can leverage the insights to make decisions like when to turn, which path to take, when to slow down or speed up.

In short, data analytics majorly represents deterministic, descriptive, and diagnostic analytics with a bit of predictive analytics. DS, on the other side, is more on predictive, prognostic, and prescriptive analytics. Machine and deep learning (ML/DL) algorithms play a very vital role here.

2.3.2. Why Data Science (DS)?

For long, we have been handling structured data and making sense out of it through BI tools. But today the scenario is totally different. We have more than 80% of multi-structured data. Data is being generated and collected from multiple sources ranging from IoT sensors, devices, and machines, business, and social applications, etc. Data formats are disparate, and data transmission protocols are also diversified. The traditional BI tools find it difficult to process such a tremendous amount of data. Thus, the multiplicity of data sources and heterogeneity of data complicate the matter further. This demands for highly sophisticated data processing tools.

Further on, if a business can understand the precise requirements of its customers from the existing data like the customer's past browsing history, purchase history, age, and income, it is definitely a boost for the business to make profits quickly and easily. Now we have competent technologies for model generation, and also there are data, which are not only vast but also varied. These transitions empower businesses to train and test prediction models efficiently. Models are also being continuously refined and readied to accurately recommend the product to their customers. Thus, model generation is being touted as one of the key differentiators of DS. Let us take weather forecasting as an example. Data from ships, aircraft, radars, satellites can be collected and analyzed to build models. These models will not only forecast the weather but also help in predicting the occurrence of any natural calamities. It will help you to take appropriate measures beforehand and save many precious lives.

In summary, BI has been an important phenomenon for achieving real business transformation. As indicated above, a perfect leverage of suitable technologies and strategies is needed for performing useful analysis on business data. This is a kind of exploratory search for uncovering actionable insights. This will answer for questions like what happened last month, why, and how it happened so, etc. It can also predict something. However, there are some critical differences between BI and DS.

Business Intelligence	**Data Science**
Uses structured data	Leverages multi-structured data (structured, semi-structured, and unstructured)
Analytical in nature. BI provides a historical report on the data	Scientific in nature. DS performs an in-depth statistical analysis on the data
Uses basic statistics with emphasis on visualization (dashboards, charts, and reports)	Leverages sophisticated statistical and does predictive analysis through machine learning (ML) algorithms
Compares historical data to current data to come out with trends	Combines historical and current data to accurately predict future performance and outcomes

2.4. THE LIFECYCLE OF A DATA SCIENCE (DS) PROJECT

A data scientist typically analyzes business data to extract meaningful insights. In other words, a data scientist solves business problems through a series of steps, including:

- Ask the right questions to understand the problem in an unambiguous manner;
- Gather data from different and distributed sources-sensor and machine data, business, web, and social application data, transactional, and operational data, personal data, etc.;
- Do the pre-processing on raw data to cleanse and convert to become compatible for easy and quick analytics;
- Feed the data into the analytic system-ML algorithm or a statistical model;
- Prepare the results and insights to share with the appropriate stakeholders.

ML and DL algorithms emerge as the most preferred way forward for data scientists to artistically shine in their obligations. Let us move over to the detail.

2.4.1. Understand the Business Domain and the Problem

The idea is to gain a deeper understanding of the problem by performing a study of the business model. The business strategy, plan, and architecture are deeply studied and well-understood before initiating the project. For example, let's say you are trying to predict the price of a 1.35-carat diamond. This phase is for understanding the terminology used in the industry and the business problem. This is for collecting a sufficient amount of relevant data about the underlying industry.

2.4.2. Data Acquisition

This involves acquiring data from different internal and external sources that can help answer business questions. Data can be extracted from different sources, including cloud IT services, business workloads running on cloud servers, scores of input/output devices including smartphones, wearables, etc., social web sites, different data stores. All kinds of data are meticulously collected, cleansed, and deposited in data repositories such as data lakes. Data brings in information.

2.4.3. Data Preparation

This is turning out to be an important factor for attaining the intended success. There may be incompatible, inconsistent, and incomplete data, and there may be some noisy data intruded into the data. Hence a data scientist has to first thoroughly examine the input data to pinpoint any gaps or data that do not add any value. This is known as data cleaning. During this process, you must go through several steps, such as data integration through data virtualization methods. Then data has to be transformed accordingly in order to facilitate the data to be used by data analytics or DS platform. ETL or ELT (extract, load, and transform) tools are the ones primed for data transformation. Data reduction is also widely recommended. There are strategies and technologies to enable data size reduction. Compression and other optimization methods are being meticulously considered to filter out repetitive, redundant, and routine data. At any cost, the reduction process should not lead to any loss or lacunae in the correctness of insights.

2.4.4. Data Exploration

Once the data is fully cleaned, real works commence with confidence. Experts are of the opinion that data scientists can perform hypothesis testing and visualize the data to understand the data better. This step is alternatively referred to as data mining, which is used to identify patterns in a data set and find important potential features with statistical analysis.

2.4.5. Model Development and Refinement

The input data is now clean and clear to be subjected to a variety of investigations. As a first step, it is essential to visualize and realize a suitable model. For example, data scientists have to understand the problem at hand is whether it is a regression or classification one. This step involves performing an exploratory data analysis in a deeper manner. Then it is to understand the relationship between variables. As indicated above, there are independent variables/features/predictors and dependent variables. Model is a sort of mathematical equation or formula capable of giving answer for any fresh set of input data. Model is then evaluated for accuracy and efficiency. Through the splitting of input data into training and test data, the obtained model can be continuously updated and upgraded to be right and relevant for giving highly accurate results.

2.4.6. Model Deployment

Once there is a predictive model in place, the next job is to deploy it in a cloud environment. Preparing, previsioning, and configuring cloud servers/ virtual machines/containers to host and run predictive models is not an easy task for data scientists. As cloud environments are steadily tending towards made of containers, the emergence of Kubernetes as container orchestration platform solution is being viewed positively. That is, we are heading towards Kubernetes-managed containerized clouds. There are a few automation tools to simplify the task of data scientists in deploying and running predictive models. Kubeflow is an open-source project that accelerates the model deployment on Kubernetes clusters.

Models are being exposed as microservices and are being containerized and stocked as container images to facilitate model portability across laptops, enterprise IT and cloud environments. There are a number of initiatives in this model deployment arena to speed up model deployment.

2.4.7. Communication

There is a need to clearly articulate the findings of data scientists to business executives. Thus, communication with clarity and confidence matters the most.

2.5. THE PROMINENT USE CASES OF DATA SCIENCE (DS)

Having understood the strategic significance of DS, business behemoths across the globe are putting DS capabilities in place in order to make sense out of data. DS or data-driven science enables better decision making, predictive analysis, and pattern discovery. It lets you:

- Find the leading cause of a problem by asking the right questions;
- Perform exploratory study on the data;
- Model the data using various algorithms;
- Communicate and visualize the results via graphs, charts, maps, dashboards, etc.

2.5.1. Airline Industry

DS is already helping the airline industry predict disruptions in travel to alleviate the pain for both airlines and passengers. With the help of DS, airlines can optimize operations in many ways. Here comes a few:

- Plan routes and decide whether to schedule direct or indirect flights;
- Build predictive analytics models to forecast flight delays;
- Offer personalized promotional offers based on customers booking patterns;
- Decide which class of planes to purchase for better overall performance.

In another example, let's say you want to buy new furniture for your office. When looking online for the best option and deal, you should answer some decision-enabling questions before making your decision.

Using this sample decision tree, you can narrow down your selection to a few websites. Ultimately it is possible to make a more informed final decision.

1. ***Healthcare:*** These companies are using DS extensively to derive actionable insights to build sophisticated medical systems to diagnose, detect, and cure diseases. There are plentiful medical sensors, instruments, and electronics devices for fulfilling a number of healthcare-related tasks. As the healthcare domain is generating a lot of patient, machine, and application data, DS capability is being leveraged to rationalize, simplify, and optimize healthcare processes.
2. ***Image Recognition:*** Still and dynamic images are being generated in large quantities these days. Cameras and smartphones are pouring out images and videos. Identifying unique patterns in images and videos, detecting objects in an image, and recognizing them are touted as the prime use cases of DS.
3. ***Recommendation and Expert Systems:*** Every complex domain is wisely and widely assisted by intelligent devices individually and collectively. Even knowledge workers in their everyday obligations are being helped with modern cognitive systems and devices in order to finish their assignments with alacrity and deftness. Netflix and Amazon give movie and product recommendations based on what you like to watch, purchase, or browse on their platforms. In the healthcare domain, there are expert systems to assist doctors, caregivers, surgeons, and nurses. There are AI-enabled chatbots in order to automate low-end repetitive tasks. There are question and answering (Q&A) systems perfectly enabled by data scientists.

4. ***Supply Chain and Logistics:*** We have discussed how DS comes handy for the airline industry in optimizing several complex things. This is expanding to the supply chain industry, and logistics companies are increasingly depending on DS to optimize routes to ensure faster delivery of products. This is resulting in increased operational efficiency.
5. ***Fraud Detection:*** This is a well-known use case of DS. Streaming analytics enhanced by DS capability is doing yeomen service for the financial sector. Banks and financial institutions use DS to pre-emptively detect fraudulent transactions.

Data is termed as the fuel for business innovation, disruption, and transformation. Data is definitely an asset for every enterprise to be right and relevant to their business partners, marketers, employees, and consumers. Every industry vertical undoubtedly generating a lot of raw data can benefit immensely through the distinct improvements and improvisations in the DS space. By incorporating DS talents, platforms, and infrastructures into their business, companies can now forecast future growth and analyze if there are any upcoming threats.

2.5.2. Prerequisites for Data Science (DS)

As articulated above, there are a number of powerful algorithms, frameworks, integrated platforms, and cloud infrastructures to augment, accelerate, and automate knowledge discovery and dissemination. Here comes a list of prerequisites:

1. ***Data Analytics Methods:*** We have been working on big, fast, and streaming data analytics. There are open source as well as commercial-grade analytics platforms for batch and real-time processing of data. As we all know, cloud computing is being pitched and positioned as the one-stop IT solution for hosting, running, and managing all kinds of customer-centric, enterprise-scale, and business-critical applications. Without an iota of doubt, cloud storages are the cost-effective, highly available, impenetrable, and universally accessible infrastructural element to stock a variety of databases, data warehouses, and data lakes.

   Most of the data analytics platforms are being already made available as cloud-based service offerings. The data transmission rate over the Internet is greatly raised through a host of patentable techniques and hence cloud environment is being seen as the way forward for all kinds of data analytics. Except extreme real-time data analytics, most of data analytics assignments are being accomplished through cloud facilities. We have analytics platform solutions such as Apache Hadoop, Spark, Flink, Storm, Samza, etc. We have HDFS as the file system. We have NoSQL databases such as HBase, Casandra, MongoDB, CouchDB, etc. We have knowledge visualization solutions such as Sisense (https://www.sisense.com/), Qlik (https://www.qlik.com/us/), Tableau (https://www.tableau.com/), etc.

2. ***AI Algorithms:*** As indicated above, the surging popularity of AI algorithms has brought in additional automation in data analytics. Predictive and prescriptive insights can be uncovered from data heaps through AI algorithms. A dazzling array of AI algorithms and approaches have rekindled a kind of interest and mystery amongst DS professionals, researchers, and other subject matter experts (SMEs) to go deep into data collections in order to emit out strategically sound insights in time. Predominantly AI comprises machine learning (ML), DL, computer vision (CV), and natural language processing (NLP). In the succeeding sections, we write about these in detail.
3. Statistics: These stays as the core of DS. The competency in statistics can help data scientists extract intelligence and obtain meaningful results tirelessly.
4. ***Programming:*** Data scientists need some programming expertise. Python is currently the leader in the DS space. R is another DS-specific language. Besides, there are enabling toolkits such as MATLAB, RStudio, and Anaconda.
5. ***Databases:*** Data warehouses and lakes are the top data storage mechanisms. A sound knowledge of database management systems (DBMS) is expected from any aspiring data scientist.

There are innovations and disruptions in the DS space. DL is a subset of ML. Feature engineering, which is manually and methodically done in ML, gets automated in DL. Thus, researchers are focusing on bringing as many automations as possible in order to lessen the load on data scientists. There are complex business and social problems that increasingly need data scientists to solve them. Also, the data size, scope, structure, and speed are varying greatly. All these increase the complexity of DS projects. Now, product vendors, cloud service providers, and AI researchers across the globe are bringing forth additional automation in the form of producing ML model engineering in an automated manner. AutoML is the new buzzword in the IT industry, gaining widespread attention. Let us now discuss the typical activities of data scientists in bringing forth trustworthy and timely insights in producing and sustaining expert, prediction, decision-making, and recommending systems.

2.6. MACHINE LEARNING (ML) ALGORITHMS

ML is touted as one of the enablers of DS. The faster maturity and stability of ML algorithms has simplified and speeded up the aspect of DS. ML algorithms have the power of empowering machines to self-learn from data. That is, identifying hidden patterns and other insights out of data is being automated through ML algorithms. ML is a kind of abstraction for data analysts.

ML models/applications sit in between datastores and users. ML models are trained using trained data and tested with test data. ML models are continuously refined in order to give the best result. Now users on giving ncw data, thc ML model comes out with a correct result. Thus, besides doing exploratory analysis, ML models result in giving predictions/ prescriptions, enabling decisions and making conclusions. Thus, ML can be called a kind of automated analytics. However, ML model creation is not an easy or straightforward task. A lot of initial work is there to arrive at competent models. That is, machines can learn automatically from data and the knowledge gained helps machines to take intelligent/context-aware decisions and plunge into the best course of actions. In other words, machines just need data and use the appropriate ML model to come out with results/outputs. That is, just input data is enough as the corresponding data processing logic comes from the embedded ML model in order to emit out output. There are several types of ML algorithms for simplifying the model creation and sustenance.

2.6.1. Supervised Learning

This is a learning in which we teach or train machines using data which is well-labeled. That means some data is already tagged with the correct answer. **For instance**, you are given a basket filled with different kinds of fruits. Now the first step is to train the machine with all different fruits one by one like this.

- If shape of object is rounded and depression at top having color Red, then it will be labeled as **Apple;**
- If shape of the object is a long curving cylinder having color Green-Yellow, then it will be labeled as **Banana.**

With this learning, we can give a new set of fruits to the machine to get the correct answer. Supervised learning classified into two categories of algorithms

- ***Classification:*** A classification problem is when the output variable is a category, such as "Red" or "blue" or "disease" and "no disease."
- ***Regression:*** A regression problem is when the output variable is a real value, such as "dollars" or "weight."

The well-known supervised learning types:

- Regression;
- Logistic regression;
- Classification;
- Naïve bayes classifiers;
- Decision trees;
- Support vector machine.

The prime advantages are:

- Supervised learning allows collecting data and produce data output from the previous experiences;
- It helps to optimize performance criteria with the help of experience.

The key disadvantages include classifying big data can be challenging and training for supervised learning needs a lot of computation time.

From predicting who is going to win the Oscars to what advertisement you are going to click and to make a prediction whether or not you're going to vote in the next election, supervised learning comes handy in answering these sorts of questions. It works because we have seen these things before. This works because we have seen the outcome/result/target and now with that learning, we can rightly predict for fresh datasets. For example, we have watched the Oscars and can find out what makes a film likely to win. We have seen advertisements before and can figure out what makes someone likely to click. We have had elections and can determine what makes someone likely to vote.

It is possible to predict Oscar winners manually by looking at the number of Oscar nominations a film receives and picking the one with the most to win. However, ML allows us to do it at a much larger scale and pick out much better predictors/independent variables or features to arrive at a best ML model. This leads to more accurate prediction, built on more subtle indicators for what is likely to happen. In supervised learning, classification is multi-dimensional in the sense that sometimes you only have two classes ("yes" or "no," or "true" or "false"). But, sometimes you have more than two. For instance, under risk management or risk modeling, you can have "low risk," "medium risk," or "high risk." Supervised ML algorithms represents a set of popular ML algorithms. This is all about predicting something. It is all about analyzing what the outcome of the process was in the past and based on that knowledge, it is to build a system that predicts the future.

2.6.2. Unsupervised Learning

Unsupervised learning is the training of machine without any information, which is neither classified nor labeled. Also, there is no guidance in unsupervised learning. Here the task of the machine is therefore to group raw and unsorted information according to hidden similarities, dissimilarities, and patterns. **For instance**, suppose a machine is given an image having both dogs and cats. The machine has no idea about the features of dogs and cat, so it can't categorize the feature in dogs and cats. But it can categorize them according to their similarities and differences. The machine is capable of discovering useful patterns and details that were previously undetected. It mainly deals with unlabeled data.

Unsupervised learning classified into two categories of algorithms:

1. ***Clustering:*** A clustering problem is for discovering the inherent groupings in the data. Customers can be grouped based on their purchasing behavior.

2. ***Association:*** This is to discover rules that describe large portions of data. The association rules are such that people that buy X also tend to buy Y.

The prominent clustering types:

- Hierarchical clustering;
- K-means clustering;
- K-NN (K nearest neighbors);
- Principal component analysis (PCA);

- Singular value decomposition;
- Independent component analysis.

In summary, unsupervised learning is empowering machines to learn without an observed outcome or target. This type of ML is less concerned about making predictions than understanding and identifying relationships or associations that might exist within the data. One common unsupervised learning technique is the K Means algorithm. This technique calculates the distance between different data points and groups similar data together. The "suggested new friends" feature in Facebook is an example of this learning. Facebook calculates the distance between users. The number of common friends for each user is calculated, and this is being used the distance metric. The more mutual friends between two users, the "closer" the distance between two users. The distance metric is used to cluster mutual friends.

While supervised and unsupervised learning work for different requirements, it's worth noting that in real-world situations, they even take place simultaneously. The most notable example is Netflix, which uses an algorithm often referred to as a recommender system to suggest new content to its viewers. You'll probably like these movies because other people that have watched these movies liked them. This is supervised learning. On the other side, these are the movies that we think are similar to other movies that you've previously enjoyed. This is unsupervised learning. Now let us move towards reinforcement learning.

2.6.3. Reinforcement Learning

What differentiates reinforcement learning from its ML brethren is the need for an active feedback loop. Whereas supervised and unsupervised learning can rely on static dataset, for example, a database and return static results. That is, the results do not change because they do not change often. Reinforcement learning actually requires a dynamic dataset that interacts with the real world. For example, think about how small kids explore the world. They might touch something hot, receive negative feedback (a burn) and eventually (hopefully) learn not to do it again. In reinforcement learning, machines learn and build models the same way. One of the well-known examples is the Deep Blue, a chess-playingchess-playing computer created by IBM. Using reinforcement learning (understanding what moves were good and which were bad), Deep Blue would play games, getting better and better after each opponent. It soon became a formidable force within the chess community and in 1996, it famously defeated chess grand champion

Garry Kasparov. There are ensemble learning methods and semi-supervised algorithms for covering up different use cases.

2.6.4. Semi-Supervised Machine Learning (ML)

This is a combination of supervised and unsupervised learning. It uses a small amount of labeled data and a large amount of unlabeled data. This combination guarantees the advantages of both while avoiding the challenges of collecting a large quantity of labeled data. These algorithms operate on data that has a few labels but is mostly unlabeled. The figure below (taken from https://towardsdatascience.com/supervised-learning-but-a-lot-better-semi-supervised-learning-a42dff534781) tells all.

This is often the case everywhere. Fortunately, semi-supervised learning algorithms operate efficiently here:

- A semi-supervised ML algorithm uses the labeled data to create a 'partially trained' model.
- Now the partially trained model labels the unlabeled data. The sample labeled data set may have many severe limitations, and hence the results of labeling are considered to be 'pseudo-labeled' data.
- Labeled and pseudo-labeled datasets are combined. This combination creates a unique algorithm that combines the descriptive and predictive aspects of supervised and unsupervised learning.

Semi-supervised learning uses the classification process to identify data assets and clustering process to group it into distinct parts. The semi-supervised GAN algorithm is a variation of the generative adversarial network (GAN) architecture to address semi-supervised learning problems. A brief on GANs.

2.6.5. Generative Adversarial Networks (GANs)

GANs are a powerful class of neural networks (NNs) that are used for unsupervised learning. GANs are generally made up of a system of two competing NN models which compete with each other and are able to analyze, capture, and copy the variations within a dataset. A generator ("the artist") learns to create images that look real, while a discriminator ("the art critic") learns to tell real images apart from fakes. That is, GANs can create images that look like photographs of human faces, even though the faces don't belong to any real person. This is pictorially illustrated below.

During training, the generator progressively becomes better at creating images that look real, while the discriminator becomes better at telling them apart. The generator tries to fool the discriminator, and the discriminator tries to keep from being fooled. The process reaches equilibrium when the discriminator can no longer distinguish real images from fakes. It has been noticed most of the mainstream neural nets can be easily fooled into misclassifying things by adding only a small amount of noise into the original data. Fortunately, the model with added noise has higher confidence in the wrong prediction than when it predicted correctly. The reason for such adversary is that most ML models generally learn from a limited amount of data. Even a slight change in a point in the feature space may result in misclassification of data.

The discriminator in a traditional GAN is trained to predict whether a given image is real (from the dataset) or fake (generated). This allows it to learn features from unlabeled images. The discriminator can then be used via transfer learning as a starting point when developing a classifier for the same dataset. Thus the supervised prediction task is to benefit from the unsupervised training of the GAN. Since most of the image features have already been learned, the training time and accuracy to perform classification will be substantially improved.

In a semi-supervised GAN, the discriminator is trained simultaneously in two modes: unsupervised and supervised:

- In *unsupervised*, the discriminator needs to differentiate between real images and generated images, like in a traditional GAN; and
- In *supervised,* the discriminator needs to classify an image into the several classes in a prediction problem, like in a standard NN classifier.

In order to train these two modes simultaneously, the discriminator must output values for 1 + *n* nodes, in which 1 represents the 'real or fake' node and *n* is the number of classes in the prediction task. In the semi-supervised GAN, the discriminator model is updated to predict K+1 classes, where K is the number of classes in the prediction problem and the additional class label is added for a new "*fake*" class. It involves directly training the discriminator model for both the unsupervised GAN task and the supervised classification task simultaneously. The entire dataset can be passed through the SGAN — when a training example has a label, the discriminator's weights are adjusted. Otherwise, the classification task is ignored, and the discriminator adjusts its weights to better distinguish between real and generated images.

The SGAN learns useful feature extractions from a very large unlabeled dataset. Also supervised learning allows the model to utilize the extracted features. This enables its classification task.

2.6.6. Why Semi-Supervised Learning?

The amount of data getting produced, collected, and stocked is growing exponentially. For instance, the number of YouTube videos getting added is simply mesmerizing. There is an exponential growth of data emanating from web and social sites, digitized entities, and connected devices. Semi-supervised learning is found to be effective in such situations. Therefore, semi-supervised learning is skillfully applied across ranging from crawling engines and content aggregation systems to image and speech recognition. The ability of semi-supervised learning to combine the overfitting and 'underfitting' tendencies of supervised and unsupervised learning creates a model that can perform classification tasks brilliantly while generalizing, given a minimal amount of labeled data and a massive amount of unlabeled data. Besides classification tasks, there exist other purposes such as enhanced clustering and anomaly detection. Semi-supervised learning is therefore the future of ML.

2.6.7. Real-World Applications of Semi-Supervised Learning

1. ***Speech Analysis:*** Since labeling of audio files is a very intensive task, semi-supervised learning is a very natural approach to solve this problem.
2. ***Internet Content Classification:* *In the Internet, there are Billions of Webpages:*** Labeling each webpage is an unfeasible process and hence semi-supervised learning algorithms are recommended to classify webpages. Even the Google search algorithm uses a variant of semi-supervised learning to rank the relevance of a webpage for a given query.
3. ***Protein Sequence Classification:*** Since DNA strands are typically very large in size, semi-supervised learning is being presented as the way forward to handle DNA data (A human DNA has approximately 3 billion base pairs) to find insights out of DNA strands.

2.7. KEY MACHINE LEARNING (ML) ALGORITHMS

2.7.1. Regression Algorithms

Regression algorithm is a supervised ML algorithm, and the output of regression is always a real or continuous value. Regression represents a statistical relationship between two or more variables in which a change in the independent variable is associated with a corresponding change in the dependent variable.

Let's say you have a website, and your revenue is based on the website traffic, and you want to predict the revenue based on site traffic. The more traffic is driven to your website, the higher your revenue would be. In a plot of revenue versus website traffic, traffic would be considered the independent variable, and revenue would be the dependent variable. The independent variable is often called the explanatory variable, and the dependent variable is called the response variable.

As website traffic increases, the revenue increases. You can draw a line to show that relationship, and then you can use that line as a predictor line. So, for example, what will revenue be if your traffic is 4,500? If you draw a perpendicular line from 4.5K on the x-axis (the traffic axis) up to the regression line/predictor line. Then you could draw another line over to the y-axis (the revenue axis) and see where it lands. You can see that when the traffic is around 4,500, the revenue is around 13,000.

Actually, there is no need to draw any line. Instead an equation/model would suffice. You could plug the independent variable into the equation to generate the dependent variable output, which is the predicted value.

2.7.1.1. Linear Regression

When there is a linear relationship between a dependent variable (which is continuous) and an independent variable (which is continuous or discrete), we would use linear regression. Linear regression answers the question, "How much?" Take a web site. As website traffic grows, how much will revenue grow? Simple linear regression considers one quantitative and independent variable X to predict the other quantitative but dependent variable Y whereas multiple linear regression considers more than one quantitative and independent variable to predict the other quantitative but dependent variable Y. Linear regression is widely used for stock market analysis, weather forecasting, and sales predictions.

2.7.1.2. Logistic Regression

Logistic regression is when the Y value on the graph is categorical (yes or no, true or false) and depends on the X variable. Whereas logistic regression predicts if something will happen or not, linear regression is generally used to predict a continuous variable, like height and weight. That is, logistic regression is used when a response variable has only two outcomes: yes or no, true or false.

Logistic regression is a binary classifier, since there are only two outcomes. Let's say you just started a company, and you are trying to figure out whether the start-up will be profitable or not. That's binary, with two possible outcomes: profitable or not profitable. So, let's use initial funding to be the independent variable. If you want to predict how much profit will be made, linear regression would be useful. For logistic regression, you will make use of a sigmoid function, and the sigmoid curve is the line of best fit. For linear regression, you would use an equation of a straight line:

$$y = b0 + b1*x,$$

where; x is the independent variable; y is the dependent variable.

Because you cannot use a linear equation for binary predictions, you need to use the sigmoid function, which is represented by the equation

$$p = 1/(1 + e - y)$$

where; e is the base of the natural logs.

Then by taking the log of both sides and solving it, you get the sigmoid function. By graphing it, you get the logistic regression line of best fit.

Another example goes like this. We have a dataset (GPAs and college ranks for several students), and we need to predict whether a candidate will get admission in the desired college or not, based on the person's GPA and college rank. Based on this labeled data, we can train the model, validate it, and then use it to predict the admission for any GPA and college rank.

Linear regression is a ML algorithm for continuous variables. However, logistic regression is a classification algorithm, not a constant variable prediction algorithm. Other supervised ML algorithms are decision trees, support vector machines (SVMs), and Naive Bayes.

2.7.1.3. Decision Trees

A decision tree is a supervised learning method used primarily for classification. The algorithm classifies the various inputs according to a specific parameter. A decision tree is a tree-shaped algorithm used to determine a course of action. Each branch of the tree represents a possible decision, occurrence, or reaction:

1. ***Root Node:*** This node represents the entire population, and this further gets subdivided into two or more homogeneous sets.
2. **Splitting:** This is a process of dividing a node into two or more sub-nodes.
3. ***Decision Node:*** When a sub-node splits into further sub-nodes, then it is called a decision node.
4. ***Leaf/Terminal Node:*** Nodes with no children (no further split) are called a leaf or terminal nodes.
5. ***Pruning:*** When we reduce the size of decision trees through node reduction (opposite of splitting), the process is called pruning.
6. ***Branch/Sub-Tree:*** A subsection of the decision tree is called a branch or sub-tree.
7. ***Parent and Child Node:*** A node, which is divided into sub-nodes, is called a parent node of sub-nodes, whereas sub-nodes are the child of parent nodes.

There are two more important concepts that you should know before implementing a decision tree algorithm: **entropy** and **information gain**. Entropy is the measure of randomness or impurity in the dataset. **Information gain** is the measure of the decrease in entropy after the dataset is split. It is also known as entropy reduction.

2.7.1.4. Support Vector Machines (SVMs)

Support vector machines (SVMs) is also a supervised learning method leveraged for classification purpose. SVMs can perform both linear and non-linear classifications. The machine learns from the existing data and predicts or makes decisions about future data. Your data set must contain known outcomes so that the machine can learn, take the data and adjust it, and apply the ML algorithm. The algorithm learns, creates a model, analyzes the model, and then uses that model to make predictions. With classification, you predict categories while in regression, and you generally predict values.

SVM is a binary classifier (a classifier used for those true/false, yes/no types of classification problems). Features are important in supervised learning. If there are several features, SVM may be the better classification algorithm choice as opposed to logistic regression. Bug detection, customer churn, stock price prediction (not the value of the stock price, but whether or not it will rise or fall), and weather prediction (sunny/not sunny; rain/no rain) are all examples.

Once you give it some inputs, the SVM algorithm will segregate and classify the data and then create the outputs. When you ingest more new data (an unknown fruit variable in this example), the algorithm will correctly classify the fruit: e.g., "apple" versus "orange." SVM is a powerful method to classify unstructured data, make reliable predictions, and reduce redundant information. SVM has applications in different areas of daily life, such as:

1. ***Face Detection:*** Using image training data, SVM classifies pixels in images like a face or non-face
2. ***Text Classification:*** Training data is used to categorize different types of documents. For instance, news articles can be classified as "business" or "entertainment."
3. ***Classifying Images:*** By classifying images with improved techniques, SVM increases search accuracy.
4. ***Bioinformatics:*** SVM algorithms have increased the effectiveness in protein homology (similarity) detection, cancer classification, gene classification, etc.

2.7.1.5. Naive Bayes

Naive Bayes is a statistical probability-based classification method best used for binary and multi-class classification problems.

2.8. ENSEMBLE LEARNING ALGORITHMS

2.8.1. Random Forest (RF)

Random forest (RF) is a popular supervised ML algorithm and this is used for both classification and regression problems. It is an ensemble learning method. The idea is to combine multiple classifiers to solve a complex problem and to also improve the performance of the model. The RF algorithm relies on multiple decision trees and accepts the results of the predictions from each tree. Based on the majority votes of predictions, it determines the final result. The classifier contains training datasets and each training dataset contains different values. Multiple decision tree models are created with the help of these datasets. Based on the output of these models, a vote is carried out to find the result with the highest frequency. A test set is evaluated based on these outputs to get the final predicted results.

An example to learn more about how a decision tree works is as follows. Suppose we want to predict whether a person will buy a phone or not based on the phone's features. For that, we can build a simple decision tree.

The parent/root node and the internal nodes represent the phone's features, while the leaf nodes are the outputs. The edges represent the connections between the nodes based on the values from the features. Based on the price, RAM, and internal storage, consumers can decide whether they want to purchase the phone. The problem with this decision tree is that we only have limited information, which may not always provide accurate results. Here, by using a RF model, it is possible to improve the results, as it provides diversity into building the model with several different features.

2.9. STEPS TO BUILD A RANDOM FOREST (RF)

- Randomly select "K" features from total "m" features where k < m;
- Among the "K" features, calculate the node "d" using the best split point;
- Split the node into daughter nodes using the best split method;
- Repeat the previous steps until you reach the "l" number of nodes;
- Build a forest by repeating all steps for "n" number times to create "n" number of trees.

After the RF trees and classifiers are created, predictions can be made using the following steps:

- Run the test data through the rules of each decision tree to predict the outcome and then store that predicted target outcome;
- Calculate the votes for each of the predicted targets;
- The most highly voted predicted target is the final prediction.

We need to understand a few different terminologies that are used in RF algorithms, such as:

1. ***Variance:*** When there is a change in the training data algorithm, this is the measure of that change.
2. ***Bagging:*** This is a variance-reducing method that trains the model based on random subsamples of training data.

3. ***Out-of-Bag (oob) Error Estimate:*** The RF classifier is trained using bootstrap aggregation, where each new tree is fit from a bootstrap sample of the training observations. The out-of-bag (oob) error is the average error for each calculation using predictions from the trees that do not contain their respective bootstrap sample. This enables the RF classifier to be adjusted and validated during training.
4. ***Information Gain and Entropy:*** These are already discussed in this chapter.
5. ***Gini Index:*** Or Gini impurity, measures the degree of probability of a particular variable being incorrectly classified when it is chosen randomly. The degree of the Gini index varies between zero and one, where zero denotes that all elements belong to a certain class or only one class exists, and one denotes that the elements are randomly distributed across various classes. A Gini index of 0.5 denotes equally distributed elements into some classes.

RF classifiers have a plethora of applications:

- In the field of banking, it is used to predict fraudulent customers.
- RFs are used to analyze the symptoms of patients and diagnose diseases.
- In the eCommerce field, recommendation lists help predict purchases based on customer activity.
- Analyze stock market trends to predict profit or loss using the RF algorithm.

2.10. TIME SERIES FORECASTING

Forecasting is a technique for making business predictions. Companies use time-series data (historical and current) to accurately forecast. This comes handy in making business decisions for the future. Time-series forecasting is the method of exploring and analyzing time-series data recorded or collected over a set period of time. Any data fit for time-series forecasting should consist of observations over a regular and continuous interval. A time series can be of annual budgets, quarterly expenses, monthly air traffic, weekly sales quantity, daily weather reports, hourly stocks' price, inbound calls per minute in a call canter) and web traffic per second. In most manufacturing companies, it drives the fundamental business planning, procurement, and production activities. Any error in the forecast will damage or degrade supply-chain efficiency. So, it's important to get the forecasts accurate in order to save on costs and is critical to success.

2.10.1. The Time Series Components

To use time-series data to develop a model, there are some patterns to be understood:

1. ***Trend:*** It represents the gradual change in the time-series data. The trend pattern depicts long-term growth or decline.
2. ***Level:*** It refers to the baseline values for the series data if it were a straight line.
3. ***Seasonality:*** It represents the short-term patterns that occur within a single unit of time and repeats indefinitely.
4. ***Noise:*** It represents irregular variations and is purely random. These fluctuations are unforeseen, unpredictable, and cannot be explained by the model.

2.11. TIME SERIES FORECASTING METHODS

1. Autoregressive Integrated Moving Average (ARIMA): The ARIMA model is a combination of the autoregressive (AR) and moving average (MR) model. The AR model forecast corresponds to a linear combination of past values of the variable. The moving average model forecast corresponds to a linear combination of past forecast errors. The "I" represents the data values that are replaced by the difference between their values and the previous values.
2. ***Seasonal Autoregressive Integrated Moving Average (SARIMA):*** This model extends the ARIMA model by adding a linear combination of seasonal past values and forecast errors.
3. **The Vector Autoregression (VAR):** This method models the next step in each time series using an AR model. The VAR model is useful when you are interested in predicting multiple time series variables using a single model.
4. ***The Long Short-Term Memory (LSTM):*** This network is a special kind of recurrent NN that deals with long-term dependencies. It can remember information from past data and is capable of learning order dependence in sequence prediction problems.

ARIMA models are classified by three factors:

- p = Number of autoregressive terms (AR);
- d = How many non-seasonal differences are needed to achieve stationarity (I);
- q = Number of lagged forecast errors in the prediction equation (MA).

For a detailed example and explanation, please visit the page (https://www.machinelearningplus.com/time-series/arima-model-time-series-forecasting-python/).

2.12. TIME SERIES FORECASTING APPLICATIONS

- Time series forecasting is used in stock price prediction to predict the closing price of the stock on each given day.
- E-Commerce and retail companies use forecasting to predict sales and units sold for different products.
- Weather prediction is another application that can be done using time series forecasting.
- It is used by government departments to predict a state's population, at any particular region, or the nation as a whole.

2.13. CLUSTERING ALGORITHMS

Clustering algorithms come under the category of unsupervised learning. Clustering algorithms work on a set of unlabeled data points and are good for grouping data points into one or more clusters based on some identified and indicated attributes. Let's say you want to travel to 20 places over a period of four days. How to visit all the places? Clustering provides the solution by grouping the places into a few clusters. The mechanism to determine these clusters is to select places that are nearest to one another. Such a grouping may result in a few clusters. Proximity is the measure used to group places within a cluster. There may be a few clusters needed to accommodate all the places. Thus, clustering comes handy in visiting all the 20 places within the allotted time.

Clustering is the method of dividing objects and then identifying and grouping specific objects together into a cluster. Thus, more than one cluster can be formed on the basis of one or more properties/attributes. If you take the trait of similarity, then all the similar objects can be clubbed together to form a cluster, whereas dissimilar objects can be joined together to form another set. Some of the well-known applications of clustering are given below:

1. ***Customer Segmentation***: Here clustering can help to answer questions as given below:
   - Which people belong to which customer segment?
   - How to group customers systematically together in forming clusters?
2. ***Social Network Analysis (SNA)***: User personas are turning out to be an excellent mechanism for clustering, which is primed for social networking analysis. We can look for similar characteristics among people and group them to form clusters accordingly.
3. ***City Planning***: This become complex when the city size soars. Therefore, mathematical, and IT solutions are mandated to simplify things. Planners need to check that an industrial zone isn't near a residential area, or that a commercial zone should not be in the middle of an industrial zone.

When choosing a clustering algorithm, the point to be noted is that whether the algorithm has the inherent capability to scale to match with the dataset, which is fast-growing. Typically, in ML problems, datasets can be in millions. The problem is that all the clustering algorithms do not scale efficiently. Many clustering algorithms work by computing the similarity between all pairs of examples. This means their runtime increases as the square of the number of examples. This is denoted as O(n2) in complexity notation. However, O(n2) algorithms are touted as impractical when the number of datasets are in millions. Therefore, the k-means clustering algorithm, whose complexity is O(n), has gained immense popularity these days. O(n) means this algorithm scales linearly with n. Here is a list of prominent clustering algorithm groups.

2.13.1. Hierarchical Clustering

This creates a tree of clusters. Without an iota of doubt, hierarchical clustering is suited to hierarchical data. Further on, any number of clusters can be chosen by cutting the tree at the right level.

2.13.2. K-Means Clustering

K-means clustering is a popular hierarchical clustering algorithm. K-means performs division of objects into K clusters that share some kinds of similarities. K, which is an integer, tells the system how many clusters have to be created. Imagine we receive data on a lot of cricket players from all over the world, which gives information on the runs scored by the player and the wickets taken by them in the last ten matches. Based on this information, we can group the data into two clusters, namely batsman and bowlers.

The steps are illustrated below:

- To begin with, we have to select a number of classes/groups to use and randomly initialize their respective center points.
- Each data point is classified by computing the distance between that point and each group center, and then classifying the point to be in the group whose center is closest to it.
- Based on these classified points, we recompute the group center by taking the mean of all the vectors in the group.
- Repeat these steps for a set number of iterations or until the group centers don't change much between iterations.

K-means has the advantage that it's pretty fast, as all we do is computing the distances between points and group centers. That is why it has achieved the minimum complexity of o(n). K-means has a couple of disadvantages. Firstly, you have to select how many groups/classes there are. This isn't always easy to calculate. K-means also starts with a random choice of cluster centers, and therefore, it may yield different clustering results on multiple runs of the algorithm. Thus, the results may not be repeatable and lack the much-needed consistency.

2.13.3. Applications of K-Means Clustering

This clustering method is widely used in a variety of real-world scenarios:

1. ***Academic Performance:*** Based on the academic marks obtained, students can be categorized into grades like A, B, or C.
2. ***Diagnostic Systems:*** The medical field increasingly uses k-means clustering algorithm towards creating intelligent medical decision support systems (DSS) and expert systems for treating and curing a variety of diseases.

3. ***Search Engines:*** When a search is performed in a search engine, the search results need to be grouped based on some considerations and the search engines have to lean upon clustering to do this difficult job effortlessly.
4. ***Wireless Sensor Networks:*** The clustering algorithm helps to pinpoint the cluster heads, each of which manages all the sensor nodes clubbed together in each cluster. There can be multiple clusters being formed out of all the sensor nodes for effectively monitoring and aggregating sensor nodes and their data.

2.13.4. Centroid-Based Clustering

Centroid-based clustering organizes the data into non-hierarchical clusters. K-means is the frequently used centroid-based clustering algorithm. This is an iterative clustering algorithm in which the clusters are formed by the closeness of data points to the *centroid* of clusters. Here, the cluster center (that is, the *centroid)* is formed such that the distance of data points is minimum with the center.

This problem is basically one of NP-hard problem, and thus solutions are commonly approximated over a number of trials. The biggest problem with this algorithm is that we need to specify K in advance. It also has problem in clustering density-based distributions. *K-means algorithm* is one of popular example of this algorithm.

2.13.5. Mean Shift Clustering

Mean shift clustering is a sliding-window-based algorithm that attempts to find dense areas of data points. It is a centroid-based algorithm. The goal is to locate the center points of each group/class, which works by updating candidates for center points to be the mean of the points within the sliding window. These candidate windows are then filtered in a post-processing stage to eliminate near-duplicates, forming the final set of center points and their corresponding groups.

2.13.6. Subspace Clustering

Subspace clustering (https://www.geeksforgeeks.org/different-types-clustering-algorithm/) is an unsupervised learning problem that aims at grouping data points into multiple clusters so that data point at single cluster lie approximately on a low-dimensional linear subspace. Subspace clustering is an extension of feature selection. The feature selection subspace clustering requires a search method and evaluation criteria. Subspace clustering limits the scope of evaluation criteria. Subspace clustering algorithm localize the search for relevant dimension and allow to them to find cluster that exist in multiple overlapping subspaces. Subspace clustering was originally purposed to solve specific computer vision (CV) problem having a union of subspace structure in the data. Professionals use this tool in social networks, movie recommendation, and biological dataset.

There are two branches of subspace clustering based on their search strategy:

- Top-down algorithms find an initial clustering in the full set of dimensions and evaluate the subspace of each cluster; and
- Bottom-up approach finds dense region in low-dimensional space and then combines to form clusters.

2.13.7. Density-Based Clustering

Density-based clustering connects areas of high density into clusters. This allows for arbitrary-shaped distributions as long as dense areas can be connected. It isolates various density regions based on different densities present in the data space. These algorithms have difficulty with data of varying densities and high dimensions. Further, by design, these algorithms do not assign outliers to clusters. Density-based spatial clustering of applications with noise (DBSCAN) is a density-based clustered algorithm.

2.13.8. Distribution-Based Clustering

This clustering approach assumes data is composed of distributions, such as **Gaussian distributions**. As the distance from the distribution's center increases, the probability that a point belongs to the distribution decreases. The bands show a decrease in probability. It is a clustering model in which we will fit the data on the probability that how it may belong to the same distribution. The grouping done maybe *Normal or Gaussian*. Gaussian distribution is more prominent where we have a fixed number of distributions and all the upcoming data is fitted into it such that the distribution of data may get maximized. This model works good on synthetic data and diversely sized clusters. But this model may have problem if the constraints are not used to limit the model's complexity. Distribution-based clustering produces clusters which assume concisely defined mathematical models underlying the data, a rather strong assumption for some data distributions. *The* *expectation-maximization (EM) algorithm*, which uses multivariate normal distributions, is one of the popular examples of this algorithm.

2.13.9. Semi-Supervised Clustering

Clustering is to identify similarities and differences between data points, but it doesn't require any given information about the relationships within the data. However, there are situations wherein some of the cluster labels, outcome variables, or information about relationships within the data are made available. Herein semi-supervised clustering is hugely beneficial. Semi-supervised clustering uses some known cluster information in order to classify other unlabeled data.

A well-known application of semi-supervised learning is a text document classifier. We all know that text documents include volumes of scripts, books, blogs, etc., which are mostly unlabeled. The power of semi-supervised learning is being felt here as it would be nearly impossible to find a large amount of labeled text documents. Precisely speaking, semi-supervised learning allows for the algorithm to learn from a small amount of labeled text documents while still classifying a large amount of unlabeled text documents in the training data.

2.14. CASE STUDY: DIABETES PREVENTION

If we can predict the onset of diabetes through the leverage of machine and DL algorithms on diabetes-associated data, it is a great help for people who are on the verge of being bracketed as diabetes patient.

- Data collection is the first and foremost activities for initiating any ML project.

The features/independent variables/decision-enabling attributes have to be identified.

For accurate diabetes prediction, the following attributes are chosen:

- **npreg:** Number of times pregnant;
- **glucose:** Plasma glucose concentration;
- **bp:** Blood pressure;
- **skin:** Triceps skinfold thickness;
- **BMI:** Body mass index;
- **ped:** Diabetes pedigree function;

- **age:** Age;
- **income:** Income.

We have all the features identified and based on that; we need to collect the data set. Going forward, all the collected data have to be cleaned to make data ready for analysis. Data cleansing is very important because there may be a number of inconsistencies, in the form of incomplete data, empty columns, outliers, incompatible data format, etc. Here is a table with organized data under different attributes. This gives a structured look indeed.

This data has a number of inconsistencies. The red color-marked fields are problematic. All the issues are fully sorted out and the table below gives a corrected look. Now we have the data set fully ready for deeper and decisive analytics. Herein, the author of this practical example has loaded the data into the analytical sandbox and applied various statistical functions on it. RStudio has functions like describe, which gives the number of missing values and unique values. The summary function gives statistical information like mean, median, range, min, and max values. The visualization techniques like histograms, line graphs, and box plots give a fair idea of the data distribution.

This problem is fit for supervised learning as all the key attributes are fully labeled. Further on, it is possible to take all the attributes consideration at one go and hence decision tree is the most appropriate one to come out with a prediction model. There are both linear and non-linear relationships. Decision tree is chosen as it is robust. That is, it allows using different combinations of attributes to make various trees and then helps to implement the one with the maximum efficiency.

Here is a decision tree.

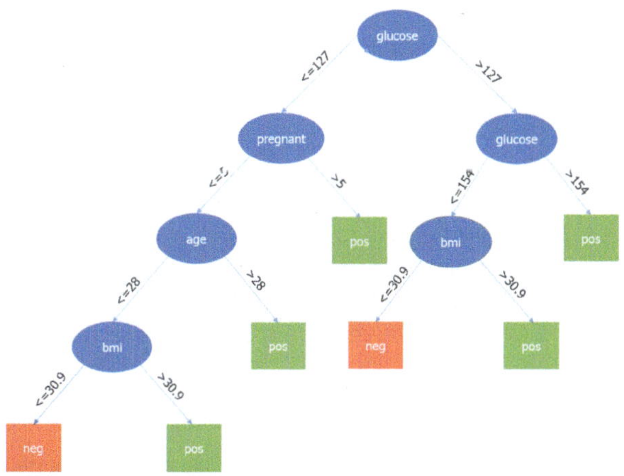

Here, the most important parameter is the level of glucose, so it is being designated as the root node. Now, the current node and its value determine the next important parameter to be considered. It goes on until we get the result in terms of pos (positive) or neg (negative).

2.15. CONCLUSION

We are experiencing big data days. Cloud-enabled data centers being set up and sustained across the globe. The computing becomes consolidated and centralized and now getting federated. With edge computing is on the anvil, computing is steadily becoming distributed. The other prominent factor is that there are competent and integrated data analytics getting deployed and maintained in cloud environments. The surging popularity of AI algorithms is another positive development. With these initiatives and implementations, the era of DS is all set to flourish. That is, data-driven insights and insights-driven decisions and actions are becoming the new normal. This chapter has detailed the various aspects associated with DS.

CHAPTER 3

PYTHON FOR DATA ANALYSIS

CONTENTS

3.1. PYTHON FOR DATA ANALYSIS

Python has gathered a lot of interest in recent times as a choice of language for data analysis.

Why Python?

- Python is open source software;
- Tremendous online society for communication;
- It is very easy to learn since it is simpler;
- It can be used as a widespread language for data science (DS) and website-based analytics product construction.

Python is an interpreted language quite compiled language. Hence it might obtain more CPU time. However, it saves programming time compared to other languages due to easy code and learning it might be a good choice.

3.2. PYTHON LIBRARIES

- Using Import:

To include the library function and modules into the python program, use the following syntax:

```
import ModuleName
```

Example:

```
import math
```

Using this math module, the functions can be invoked using:

```
math.functionName
```

Example:

```
math.sqrt(25)
```

We can define an alias to the module to simplify:

```
Import ModuleName as AliasName
```

Example:

```
import math as m
print(m.sqrt(25))
```

- Using From…. Import:

The modules can also be imported using the keyword from:

Syntax:

```
from Module import *
```

Example:

```
from math import *
print(math.sqrt(25))
```

For scientific computations and data analysis, we need of the following list of libraries.

1. **NumPy:** It is a Numerical Python. The most use of Numpy is n-dimensional array. It contains basic Fourier transforms, all linear algebra functions, advanced random number capabilities and integration tools with other low-level languages like FORTRAN, C, and C++
2. **SciPy:** It is Scientific Python. It is built on NumPy. For high-level science coding and for engineering modules, i.e., Fourier transform, linear algebra, and sparse matrix, it is a very useful library function.
3. **Matplotlib:** Using the Matplotlib enormous variety of graphs, starting from histograms to line plots to heat plots can be plotted. We can use Pylab features in ipython notebook. Without inline the ipython environment is similar to the MATLAB environment.
4. **Pandas:** These are very useful for structured data operations and manipulations. It is widely used for data administration and preparation. Pandas were added relatively recently to Python and have been instrumental in boosting Python's usage in the data scientist community.
5. **Scikit:** It is used to Learn for ML. It is Built on SciPy, NumPy, and matplotlib, this library contains a lot of well-organized tools for ML and statistical modeling, including regression, clustering, classification, and dimensionality reduction.
6. **Statsmodels:** It is used for statistical modeling. Stats models is a Python module that allows users to explore data, estimate statistical models, and perform statistical tests. A widespread list

of evocative statistics, statistical tests, plotting functions, and result statistics are offered for different types of data and each estimator.

7. **Scrapy:** It is used for web crawling. It is a constructive framework for getting precise patterns of data. It has the competence to start at a website home URL and then dig from end to end web pages contained by the website to collect information.
8. **SymPy:** It is used for figurative computation. It has comprehensive capabilities from basic symbolic arithmetic to algebra, calculus, quantum physics, and discrete mathematics.
9. **Requests:** It is used to access the web. It works equivalent to the standard python library urllib2 but is a lot easier to code. You will find subtle differences with urllib2, but for beginners, Requests might be more expedient.

Supplementary libraries, we may need:

- **OS:** This library for file operations and operating system.
- **networkx and igraph:** It is used for graph-based data manipulations.
- **Regular Expressions:** It is used for finding patterns in text data.
- **BeautifulSoup:** This library is used for scrapping web. It is inferior to Scrapy as it will haul out information from just a single webpage in a run.

3.3. SCIENTIFIC LIBRARIES IN PYTHON-NUMPY, SCIPY, MATPLOTLIB, AND PANDAS

NumPy is an open-source Python package. NumPy is a 'Numerical Python.' It is the collections of multidimensional array objects and a compilation of routines for processing of array.Jim Hugunin developed NumPy which is the ancestor of Numeric.

3.3.1. Operations Using NumPy

Using NumPy, a developer can perform the following operations on arrays:

- Mathematical operations;
- Logical operations;

- Fourier transforms and routines which are used for shape management;
- Linear algebra operations.

3.3.2. NumPy for MATLAB

NumPy is frequently used on the packages like Scientific Python and plotting library Mat–plotlib. This amalgamation is broadly used as an alternate for MATLAB, a popular stand for technical computing. Though, Python substitute to MATLAB is now seen as a more contemporary and inclusive programming language.

- **Installation of NumPy:** The Numpy can be installed through python package installer PIP Command(Windows):

  Python-m pip install NumPy

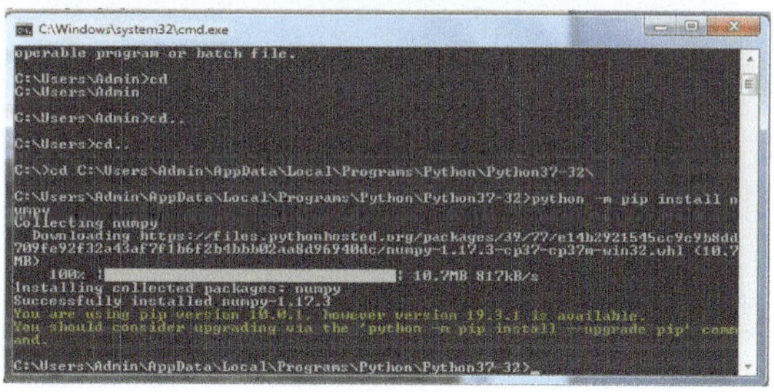

- To install SciPy:

  Python-m pip install SciPy

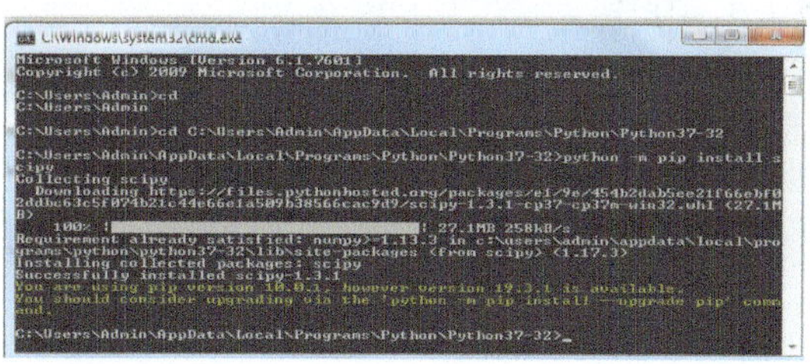

- To install matplot:

  Python-m pip install matplot

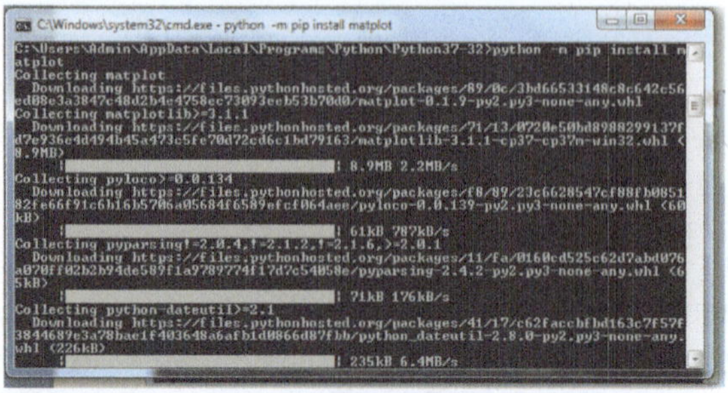

- To check the version of SciPy:

  scipy.version.version

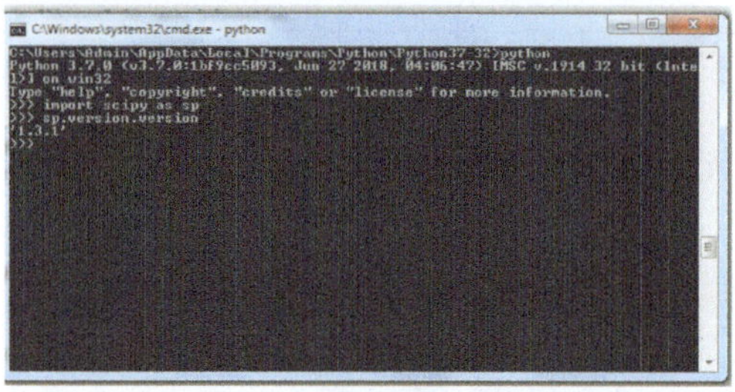

Let's plot a simple function with *Matplotlib*. First, we'll import *SciPy* and *Matplotlib* and then we can define some points on the (0, 1) interval with:

```
import scipy as mp                  #import SciPy
import matplotlib.pylab as pt       #import matplot
tt = mp.linspace(0, 1, 100)         #defining points
pt.plot(tt, tt**2)                  #To plot a parabola
pt.show()                           #To show the output
```

We can see the parabola like:

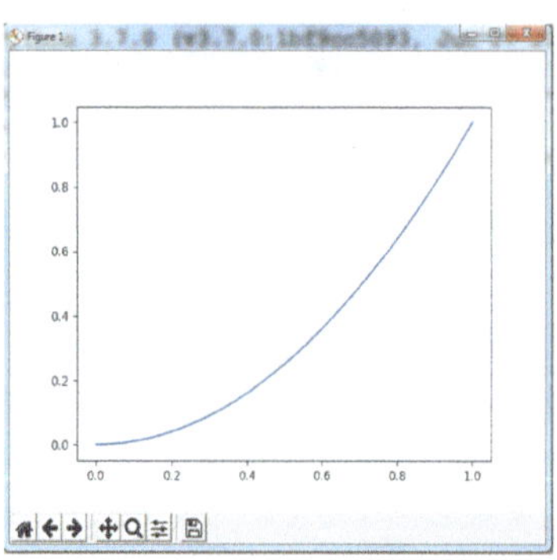

3.3.3. Sine Wave Plot Using Matplotlib

The sine wave can be produced using the matplotlib module. The function pyplot is used to draw the sine wave, and the values can be define using the NumPy module. The NumPy is the library which is the collection of the same data items. The arrange function is used to create the sequence of values which have even distance. This arrangement can be used with the following syntax:

Numpy.arange (start value, stop value, step value, datatype)

Here the stop value is a must, and the remaining parameters are optional.

Sample program:

```
import NumPy as n
import matplotlib.pyplot as plott
xx = n.arange(0, 4 * n.pi, 0.1)
yy = n.sin(xx)
plott.title("sine wave form")
plott.plot(xx, yy)
plott.show()
```

Output:

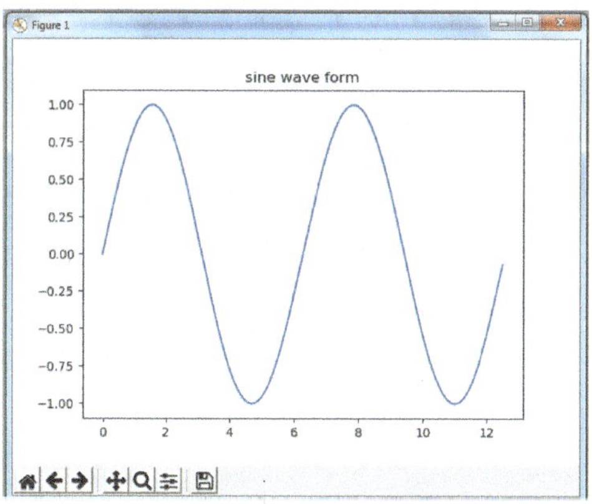

3.3.4. Drawing 3D Plot Using Matplotlib

To draw 3D images, the function axes3d has to be imported from the module mpl_toolkits.mplot3d. The figure function is used to draw the canvas.

Syntax:

```
plot_wireframe(self, X axis, Y axis, Z axis, *args, **kwargs)
```

Program:

```
from mpl_toolkits.mplot3d import axes3d
import matplotlib.pyplot as pt
chart = pt.figure()
chart3d = chart.add_subplot(111, projection='3d')
X, Y, Z = axes3d.get_test_data(0.05)
chart3d.plot_wireframe(X, Y, Z, color='g,' rstride=20, cstride=30)
pt.show()
```

In this above code, X, Y, Z is the input 2D-array data, stride is the step size. Downsampling stride is in every direction. These arguments are normally restricted with rcount and ccount. If only one of rstride or cstride is set, the other defaults to 1. Assigning stride to zero causes the data to be not sampled in the subsequent direction, producing a 3D line plot rather than a wireframe plot.

Output:

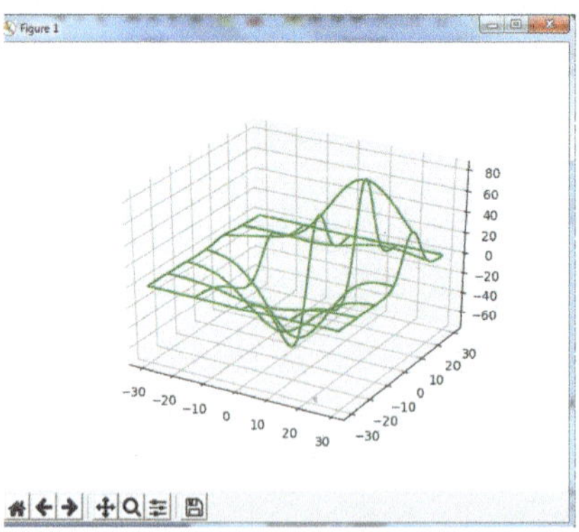

3.3.5. Data Analysis Using Pandas

Pandas is the mainly trendy python library that is used for data analysis. It provides tremendously optimized performance with back-end source code is exclusively written in C or Python.

Installation of Pandas:

Pandas is installed via pip

Python-m pip install pandas

To import panda library:

import pandas as pand

Using panda data analysis can be done on:

- Series
- Data Frames

Create series with data:

s = pand.Series(data)

Here, data is:

- A scalar value which may be integer value, string.
- A python dictionary which is a combination of key, value pair.
- A N-d array.

Create series with data, and index:

si = pand.Series(Data, index = Index)

- *Example 1:* When data contains the scalar values.
  import pandas as pand
  data1 =[1, 3, 4, 5, 6, 2, 9]
  s = pand.Series(data1)

```
Index1 =['a,' 'b,' 'c,' 'd,' 'e,' 'f,' 'g']
si = pand.Series(data1, Index1)
print(s)
print(si)
```

Output:

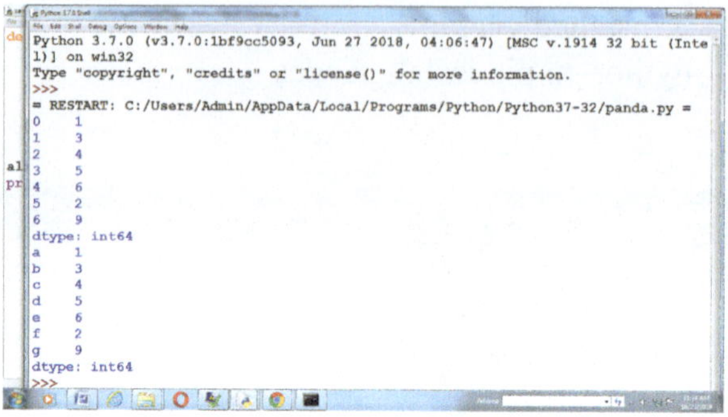

- *Example 2:* When data contains dictionary.

```
import pandas as pd
dictionary1 ={'a':1, 'b':2, 'c':3, 'd':4, 'e':5}
sdic= pd.Series(dictionary1)
print(sdic)
```

Output:

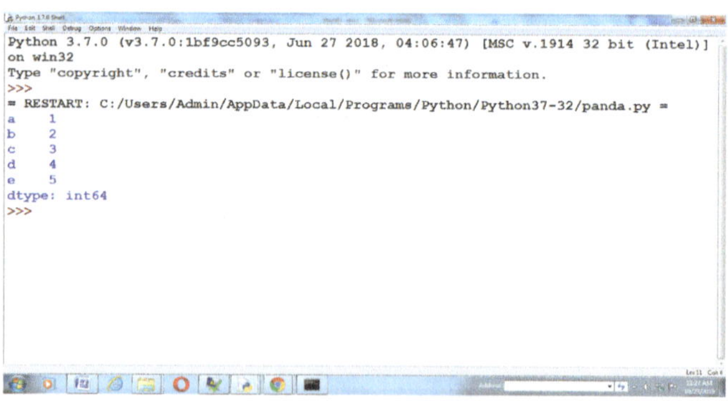

- *Example 3:* When data contains Ndarray.

```
import pandas as pand
Data =[[20, 30, 40], [50, 60, 70]]
sarr = pand.Series(Data)
print(sarr)
```

Output:

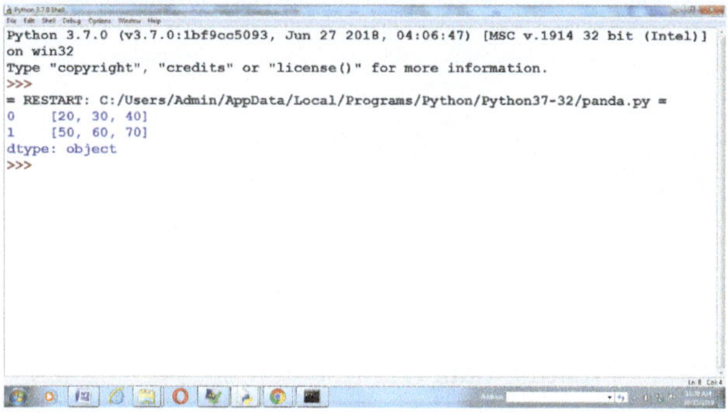

3.3.6. Create Dataframes

Data frames are the data structures which is declared as the rows and columns. It can be 2D arrays.

```
import pandas as pand
df = pand.dataFrame(data)
```

Here, data can be:

- Dictionaries
- Series
- 2D-numpy Ndarray
- *Example 1:* Dataframe with dictionaries

```
import pandas as pd
dict1 ={'a':1, 'b':2, 'c':3, 'd':4}
dict2 ={'a':5, 'b':6, 'c':7, 'd':8, 'e':9}
Data = {'first':dict1, 'second':dict2}
df = pd.DataFrame(Data)
print(df)
```

Output:

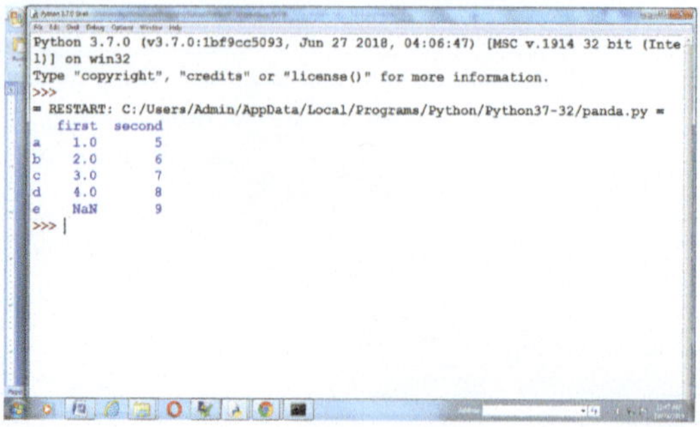

- *Example 2:* Creating data frame when data is a series

```
import pandas as pand
s1 = pand.Series([1, 3, 4, 5, 6, 2, 9])
s2 = pand.Series([1.1, 3.5, 4.7, 5.8, 2.9, 9.3])
s3 = pand.Series(['a,' 'b,' 'c,' 'd,' 'e'])
Data1 ={'first':s1, 'second':s2, 'third':s3}
datseries = pand.DataFrame(Data1)
print(datseries)
```

Output:

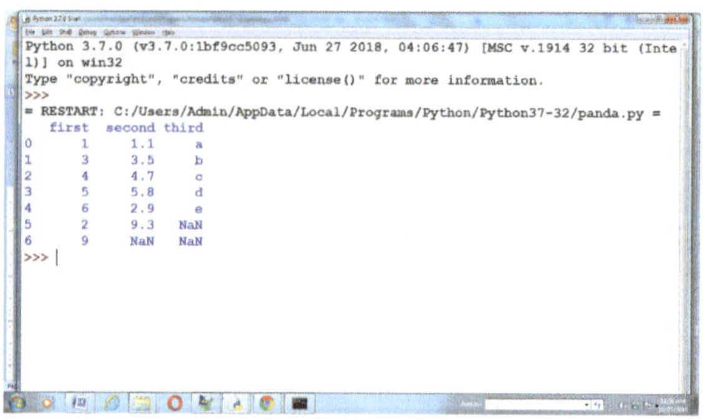

- *Example 3:* When data is 2D-numpy Ndarray

```
import pandas as pand
d1 =[[2, 3, 4], [5, 6, 7]]
d2 =[[2, 4, 8], [1, 3, 9]]
Data1 ={'first': d1, 'second': d2}
df2d = pand.DataFrame(Data1)
print(df2d)
```

Output:

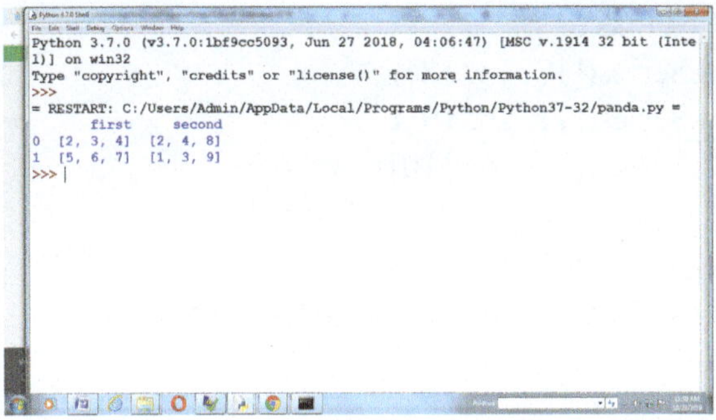

3.3.7. Using CSV Files

Another way to create a DataFrame is by importing a csv file using Pandas.

```
Create a CSV file with the name cars.csv
import pandas as pand
vehile = pand.read_csv('cars.csv')
print(vehile)
```

Output:

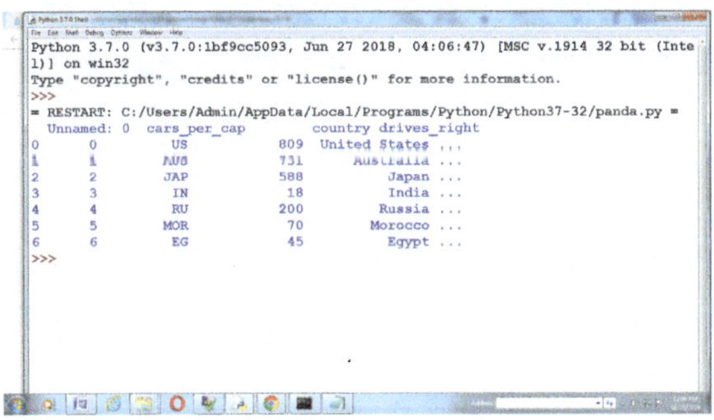

1. Store DataFrame in CSV file format:

We can store the created dataframe into the csv file using the following syntax:

to.csv('filename,' index = "False|True")
filename-name of the csv file
index-True if the default value to be overwritten
index-False if the default value should not be overwritten
Default index value is True

Example:

```
import pandas as pd
s1 = pd.Series([10, 14, 18])
s2 = pd.Series([11, 15, 19])
s3 = pd.Series([12, 16, 110])
dframe = pd.DataFrame([s1, s2, s3])
dframe.columns =['first,' 'second,' 'third']
dframe.to_csv('sample.csv,' index = False)
dframe.to_csv('sample1.csv,' index = True)
```

Output:

Sample.csv

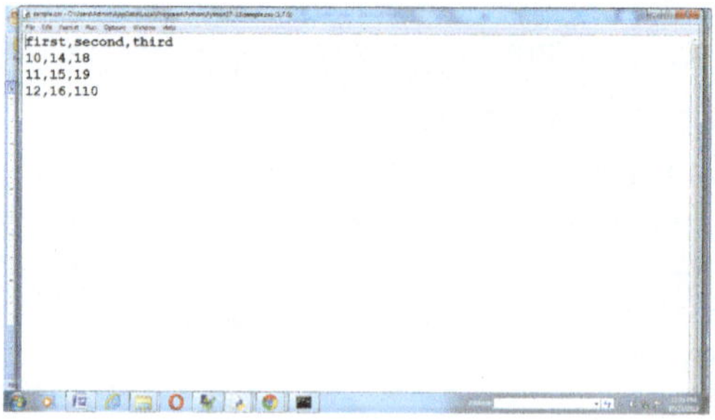

Sample1.csv

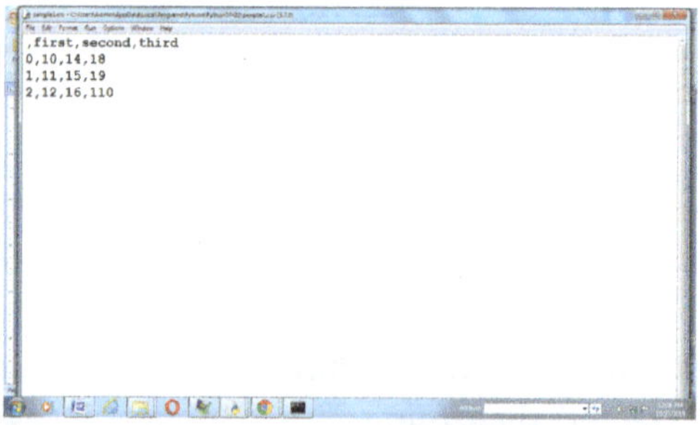

3.3.8. Data Visualization Using Matplotlib

The matplotlib is used to display the bar chart, line chart, and scatterplots, which can be produced from the data analysis.

Include matplotlib in a file using the keyword import

Example: From matplotlib import pyplot as plt

```
x = [1950, 1960, 1970, 1980, 1990, 2000, 2010]
y = [300.2, 543.3, 1075.9, 2862.5, 5979.6, 10289.7, 14958.3]
plt.plot(x,y,color='red,'marker='o,'linestyle='solid')
plt.title("X axis")
plt.ylabel("y axis")
plt.show()
```

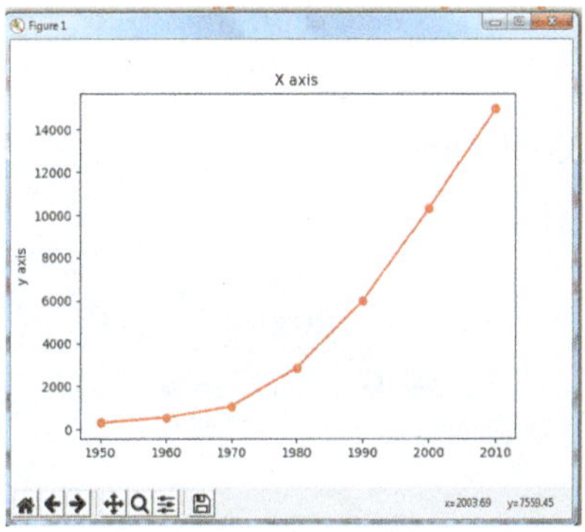

3.4. MACHINE LEARNING (ML)

3.4.1. Introduction

It will be more comfortable and essential if we have one language to understand and sense the data just like the way of human beings do. In another word, the language that is needed to take out the patterns from the raw facts using some artificial Intelligence (AI) is much needed. For this, the algorithm or methods can be used to haul out the data. The solution for this prerequisite from the computer science is the ML. This machine learning makes the machine to analyze and be trained from the periodic learning from the experience without the human interference.

3.4.2. Prerequisite

Before getting started with ML, the programer need to have some basic knowledge in terms of AI for learning machines. To make the machine to think and learn, the knowledge in Python and NumPy, SciPy, Matplotlib, Scikit-learn also be needed.

3.4.3. Machine Learning (ML) Model

Professor Mitchell defined the ML as:

> *"A computer program is said to learn from experience E with respect to some class of tasks T and performance measure P, if its performance at tasks in T, as measured by P, improves with experience E."*

In simple words, the ML is based on the three important parameters,

- Tack (T);
- Performance (P); and
- Experience (E).

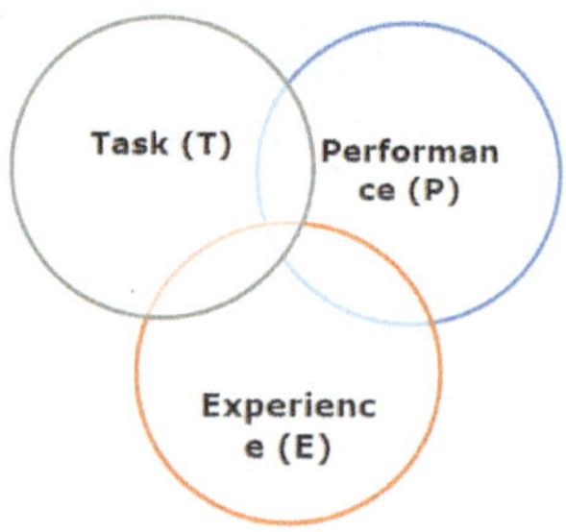

ML is the part of AI with learning algorithms and methods:

- Develop and improve performance (P);
- At carry out some task (T);
- Over time with knowledge experience (E).

3.4.3.1. Task (T)

Task is the solution to the real world problem.

Example: Out of many kind of loan strategies from different bank, getting the best and economic housing loan. From the conventional programming approach, it is very difficult to solve ML-based tasks. Regression, clustering, transcription, and structured annotations are the best examples of ML-based tasks to be solved.

3.4.3.2. Experience (E)

Human being is experienced by learning from the various situations and environment, relationship among different people. As it is, experience can be gained from the previous execution of an algorithm and the collected data. To gain this knowledge, the dataset will be executed iteratively.

There are some ways to gain experience through:

- Supervised learning;
- Unsupervised learning;
- Reinforcement learning.

Based on the above the algorithm is classified as:

- Supervised learning algorithm;
- Unsupervised learning algorithm;
- Reinforcement learning algorithm;
- Semi-supervised learning algorithm.

Supervised learning is based on the real word problem and the solution based. For example, movies, product sales. It is classified into two types.

1. **Regression:** Response from the continuous values
2. ***Classification:*** Based on analyzing and classifying. For example, Male, Female, and positive and negative comments

Unsupervised learning is based on the fraud detection and fault finding. It works on the no data label. That is unlabeled data sets.

Reinforcement learning is the process of giving the feedback to the system so that the system can adjust dynamically and works perfectly.

3.4.3.3. Performance (P)

Performance is measured by how much the machine is performing as per requirement and expected time bounce. It is a quantitative metric which measures the performance of tasks T from the Experience (E).

3.4.4. Python for Machine Learning (ML)

For ML, make sure the below packages are installed in your machine.

3.4.4.1. Download and Install SciPy libraries

The most important SciPy libraries to be present are:

- SciPy;
- matplotlib;
- NumPy;
- sklearn;
- pandas.

Installation of SciPy, matplotlib, NumPy, and pandas we have seen in the deep learning (DL) chapters. Now let us see how to install sklearn.

The scikit can be installed using the following command:

Python-m pip install-U scikit-learn

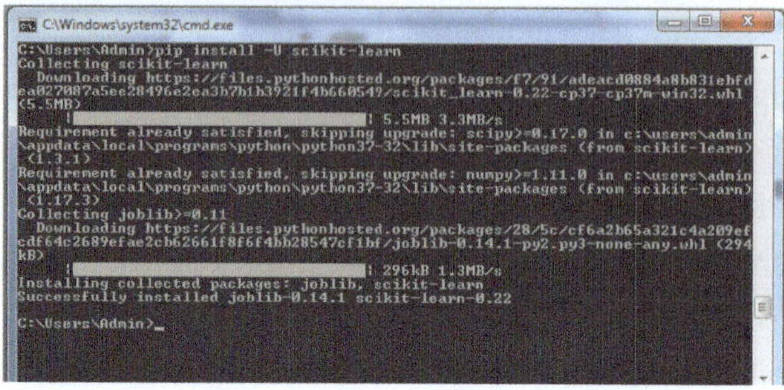

To verify the installation of scikit run the command:

Python-m pip show scikit-learn

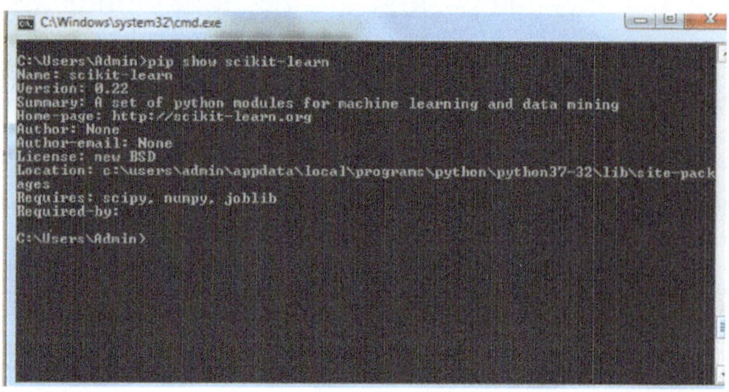

To install sklearn:

Python-m Pip install sklearn

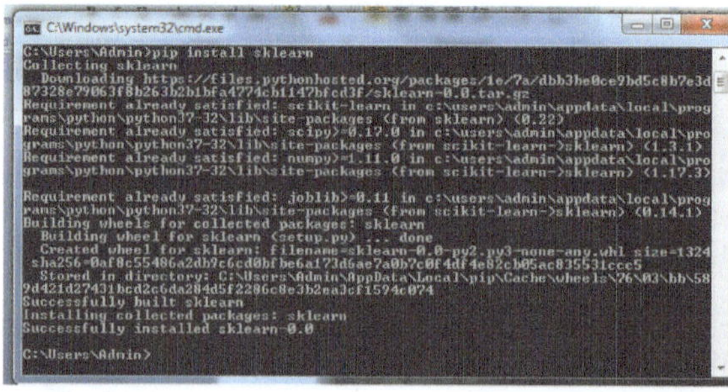

Writing a Python program using Anaconda Prompt or terminal:

1. Open the anaconda prompt:

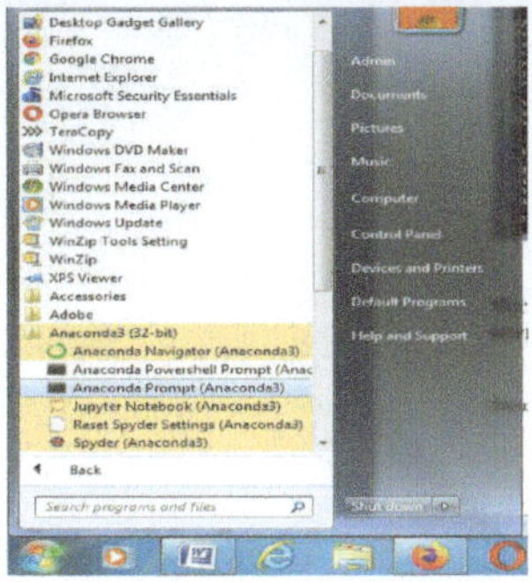

2. Enter into python prompt by giving the command:

Python

3. Check for the libraries whether they have installed:

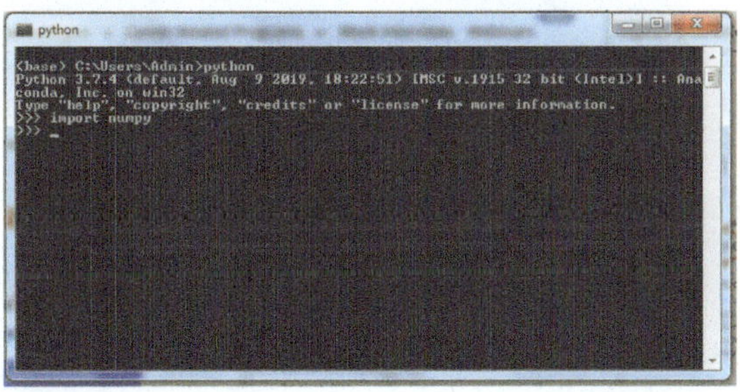

Getting no error is confirming that the libraries are installed properly.

4. Open the Anaconda Navigator

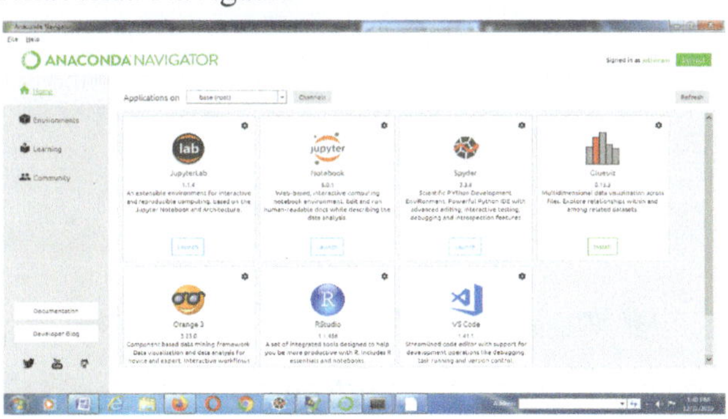

5. Launch the Spyder and open:

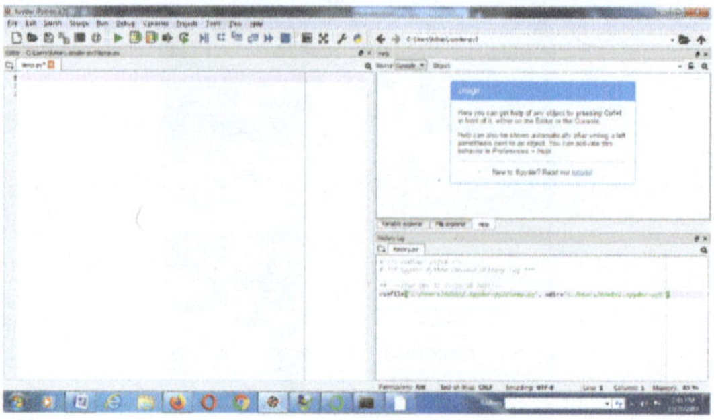

6. Now type:

Print("Hello Python")

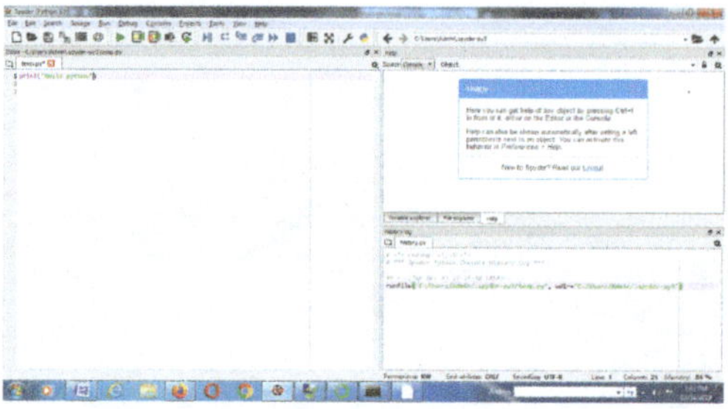

7. Run the triangle symbol to execute the console:

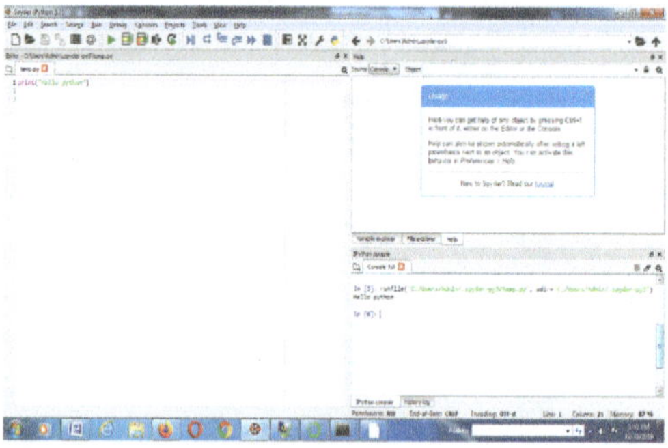

3.4.4.2. Preparing Dataset for Python

To prepare a dataset for python, we need to ensure the following steps:

1. Create or get the dataset: The CSV file can be created in Microsoft Excel and it can be saved as a CSV file.

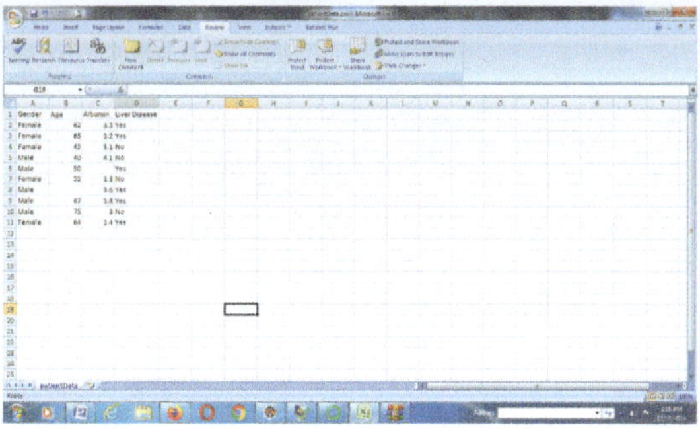

Here we have created the patientData.CSV file.

2. Import all the necessary libraries by using the following code:

   import NumPy as np1

   import matplotlib.pyplot as plt1

   import pandas as pd1

   Run the code to check for the confirmation of library installation:

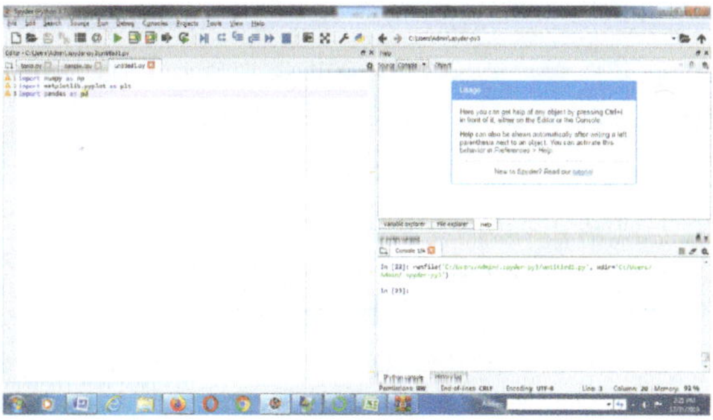

Since all the libraries installed successfully, there is no error.

3. Handling of missing data in dataset: When we are working with the dataset always there will be a problem of missing data in the given dataset. To run the code effectively, the missing data has to be filled very carefully. The syntax to read the data from the CSV file is as follows.

   pandas.read_csv('filename.csv')

   By using the above syntax the code is altered as:

   import NumPy as np1
   import matplotlib.pyplot as plt1
   import pandas as pd1
   data = pd1.read_csv('patientData.csv')

While executing this dataset we are getting the following output as:

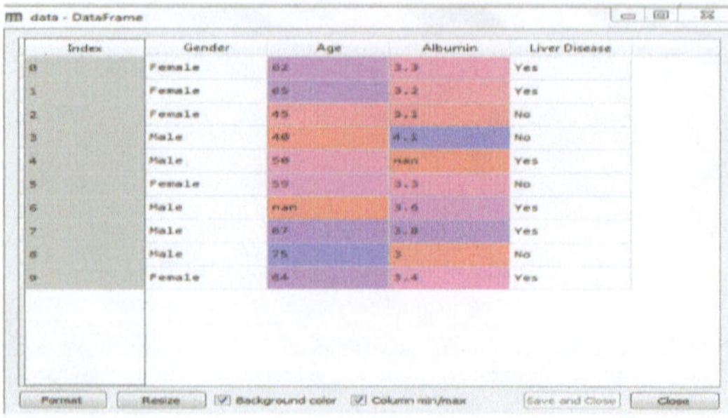

Index	Gender	Age	Albumin	Liver Disease
0	Female	62	3.3	Yes
1	Female	65	3.2	Yes
2	Female	45	3.1	No
3	Male	48	4.1	No
4	Male	58	nan	Yes
5	Female	59	3.3	No
6	Male	nan	3.6	Yes
7	Male	67	3.8	Yes
8	Male	25	3	No
9	Female	64	3.4	Yes

Here some of the rows contain the value 'nan.' That is Not Any Number. It is indicating that there is no data or the data is missing.

The missing data can be filled with the "MEAN" value.

```
XX = data.iloc[:,:–1].values

YY = data.iloc[:, 3].values
```

Here XX = data.iloc[:,:–1].values is used to extract the index starting from beginning to the end except the last rows. LOC and ILOC are the functions to retrieve the data from the row of the dataset.

To handle the missing data from the dataset we need SKLEARN, hence import the sklearn. To fill the missing data, we have certain algorithms called multivariate imputation algorithms. It uses the complete set of obtainable characteristic dimensions to guess the missing values of the dataset. This algorithm is inside the sklearn package and it can be imported by:

```
from sklearn.preprocessing import Imputer
```

Now alter the code,

```
imputer = Imputer(missing_values = 'NaN,' strategy = 'mean,' axis = 0)
imputer = imputer.fit(XX[:, 1:3])
XX[:, 1:3] = imputer.transform(XX[:, 1:3])
```

Now all the missing value will be filled approximately by using the function mean.

3.4.5. Plotting Graphs

For visualization of data frames, we need the library seaborn. This is used to visualizes the statistical data, which provides an attractive way of visualizing the data. The seaborn is based on the matplotlib.

We can draw or plot the different types of graphs from the dataset. One type is the BOX plot. It can be drawn by the following code:

```
import pandas as pd1
import NumPy as np1
import matplotlib.pyplot as plt1
import seaborn as sns1
dfl = pd1.read_csv('patientData.csv')
sns1.boxplot(x=dfl['Age'])
sns1.boxplot(x=dfl['Albumin'])
```

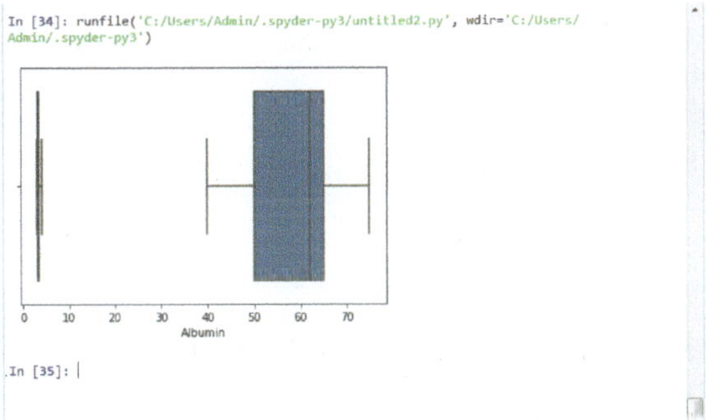

To count the frequency of the data we may use a counterplot.

For counter plot:

```
sns.countplot(x='X axis data,'hue='value,'data=dataframe)
```

The code for the counter plot:

```
import pandas as pd1
import NumPy as np1
import matplotlib.pyplot as plt1
import seaborn as sns1
df = pd1.read_csv('patientData.csv')
sns1.countplot(x='Age,'hue='Liver Disease,'data=df)
```

```
In [35]: runfile('C:/Users/Admin/.spyder-py3/untitled2.py', wdir='C:/Users/
Admin/.spyder-py3')
```

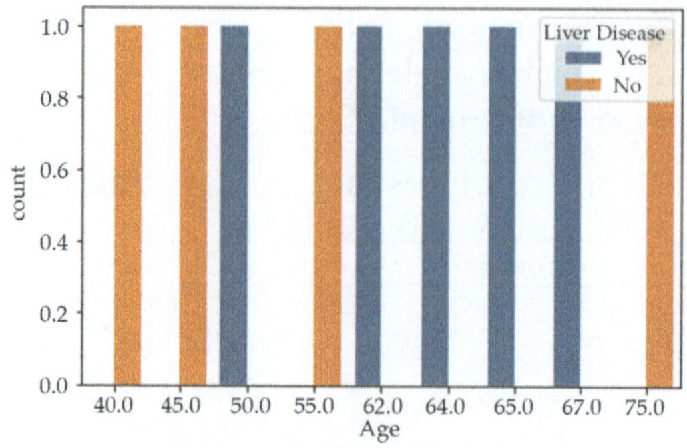

3.4.6. Categorization of Machine Learning (ML)

Base on the problem set, there are three types of ML:

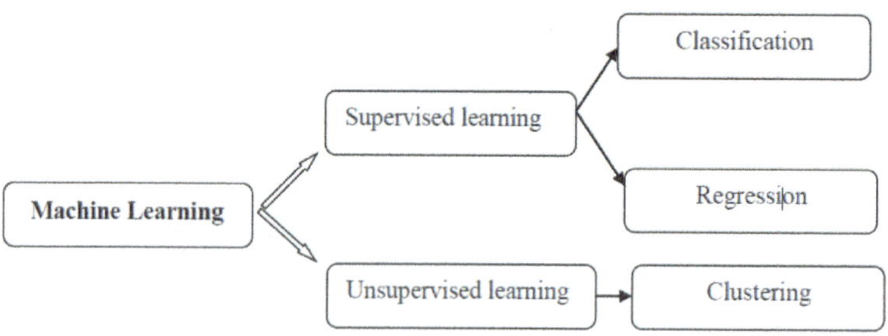

1. ***Supervised Learning:*** The labeled data set is the training data for the supervised learning. The knowledge and the relationship between the feature and label set can be learnt from the labeled data set. Consider feature vector x and the label YЄL, where L = {L1, L2,……..Lc} c-the value takes from 2 to 100. It is called classification.

   If the feature vector X is corresponds to the real vector R, then YЄR, is called regression. The knowledge which id obtained from the supervised learning is often used for the prediction and recognition the problems.

2. **Unsupervised Learning:** The unlabeled dataset is the training data for unsupervised learning. It is used in Probability density estimation, Clustering, Dimensionality reduction, finding association among features.

3. ***Reinforcement Learning:*** The reinforcement algorithm is mostly used in decision making. The applications of reinforcement include automatic chess player, automatic vehicle driving, etc.,

3.4.7. Machine Learning (ML) Algorithms

1. Linear regression;
2. Logistic regression;
3. Decision tree;
4. SVM;
5. Naive bayes;
6. kNN;
7. K-means;
8. Random forest (RF);
9. Dimensionality reduction algorithms;
10. Gradient boosting algorithms:
    i. GBM;
    ii. XGBoost;
    iii. LightGBM;
    iv. CatBoost.

3.4.7.1. Linear Regression

The real values are estimated by linear regression. The real values are the number of calls, total sales of the product and the cost of houses, etc. It is estimated based on the continuous values. The dependent and the independent values are related by setting the lines. The setting of this fit line is called regression line. The regression line is represented by:

$$Y = a + bX$$

In this equation:

- 'Y' is the dependent variable;
- 'a' is an intercept;
- 'X' is an independent variable;
- 'b' is a slope.

Here a and b are the coefficient which is derived from the sum of square of difference between data point ad regression line.

There are two types of linear regression. They are:

1. linear regression; and
2. multiple linear regressions.

Take X as a feature vector and the Y as a response vector:

$X = [x_1, x_2, x_3 \ldots\ldots x_n]$

$Y = [y_1, y_2, y_3 \ldots\ldots y_n]$ …for the n observation, here we take n = 10

X	0	1	2	3	4	5	6	7	8	9
Y	1	3	2	5	7	8	8	9	10	12

The scatter plot can be drawn for the above as:

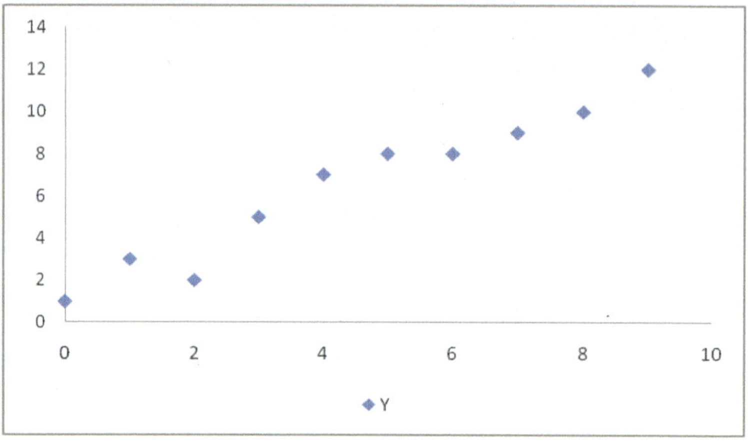

Now for the above, we have to find the best fit that is the regression line, which will enable us to find the response for any newly added features.

Let's take h(xi) is the predictable response value.

B0 and B1 is taken as the coefficient values.

The equation for finding the regression line is as follows:

There may be a residual error represented as Ei can be added to the above equation. But we have to try to reduce the residual error.

It can be:

h(xi) + Ei

Now the Ei can be found as Ei=yi-h(xi)

3.4.7.2. SVM

A Support Vector Machine, it is a model for supervised learning which is using the classification algorithms to two-group classification problems. After the training data set, we can able to categories the new types of texts.

It takes the data as the input and produces the output in two dimensions called hyperplane which separate tag. This line is called decision boundary.

Sample:

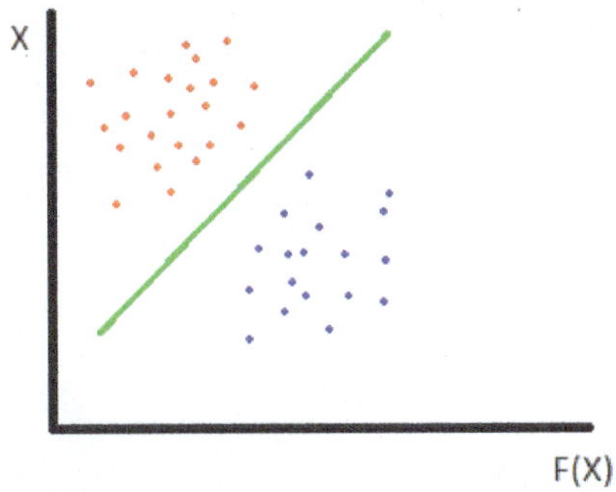

3.5. MACHINE LEARNING (ML) WITH INTERNET OF THINGS (IOT)

The recent applications works smarter than before with the modern technologies like the internet of Things (IoT) which gives exploitation among the modern world. Even though there are development and enhancement in technologies, the data pool becomes larger and larger, which urges the new techniques called ML to analyze and extract the data from the collection. The ML techniques is collaborated with IoT to make the technology automatic one. It may include smart home, smart lighting systems, automation, etc.

3.5.1. Internet of Things (IoT)

The purpose of IoT to minimize the cost of energy and saves them money and time by developing the smarter environments, smart homes and smart cities. Many of the industries has reduced their cost by applying the IoT. Researchers are working on IoT for advancement which is in trending. The data can be passed among the different devices to obtain their performance and this transferring of data is automatic without any human interference and also it doesn't require any input.

There are four very basic things associated with the IoT. They are:

- sensors to sense the environment;
- networks for processing;
- data analysis; and
- system monitoring.

A variety of technologies are incorporated into IoT and these technologies need to be connected along with necessary conditions. For this, a suitable protocol should be developed for the enhanced communication among a variety of things. There are different types of communication protocols which is mainly divided into three major categories:

1. ***Device to Device (D2D):*** This type of protocols is mainly used to establish the communication in between the mobile phones which are nearby. It is known as the next generation of the mobile phone network.

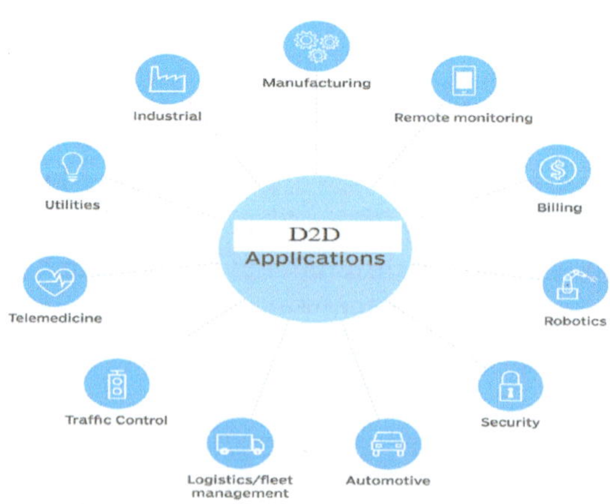

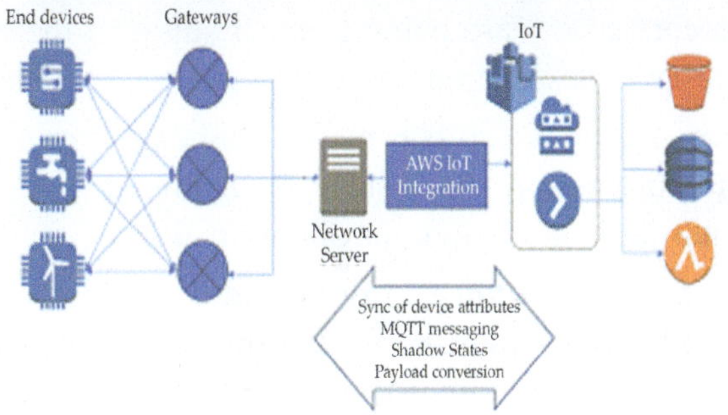

2. ***Device to Server (D2S):*** It is used to collect the information from the devices and being sent to the server for the storage. The server may be near or far. An example for this types of protocols are cloud processing.
3. **Server to Server (S2S):** The data is transmitted from one server to another server. This protocol is applied in the cellular types of networks. The data preparation and the processing is decisive challenge. For this challenge edge analytics, stream analysis, IoT analysis on database can be applied. Among the mentioned types, it can be decided based on the type of application.

For data processing and data preparation before transmitting data to another application, there are two analytical methods are used.

They are:

- Fog;
- Cloud processing.

In simple words the task of IoT is:

- The data has been collected through the sensors and IoT devices from the environment;
- From the raw data, knowledge is been extracted;
- The data is ready to transferred.

3.6. MACHINE LEARNING (ML) APPLICATION WITH IOT

- ***Energy:*** It is used to reduce the cost of the energy being used with the help of Arduino Mega. Example Coffee Machine, air conditioners, and Light that are connected to IoT.
- ***Routing Traffic:*** For /traffic routing different ML algorithms and

different sensors are being used. It suggests the different routed to the same destination.

- ***Home Automation:*** The major application of IoT proposed to home automation. They have been implemented in apartments for humidity and light control application.
- ***Industry Automation:*** In organizations and several companies, the applications are used for the traffic management, health care and manufacturing industries. The human resource and time can be saved with the help of IoT applications.

3.7. ALGORITHM

Definition: An algorithm is a sequence of finite number of steps to solve the problem. An algorithm contains the specific rules which is to be followed while writing.

Example algorithm:

- Start;
- Read the value of radius r;
- Calculate Areaofcircle=3.14*radius*radius;
- Print the result as Area of the circle;
- Stop.

Characteristic of algorithm:

- An algorithm should be necessarily specific and definite;
- No repetition of algorithm instruction is permitted;
- The algorithm should be terminated at the end followed by the attainment of the result;
- This should be written in a sequential pattern.

Qualities of a good algorithm:

- **Time:** Lesser time required.
- **Memory:** Less memory required.
- **Accuracy:** Suitable or correct solution obtained.
- **Sequence:** Must be sequence and some instruction could be repetitive until certain requirement is met.
- **Generability:** Used to solve a single problem and a certain range of input data could also be handled by the algorithm.

3.8. BUILDING BLOCKS OF ALGORITHMS (INSTRUCTIONS/STATEMENTS, STATE, CONTROL FLOW, FUNCTIONS)

Any algorithm can be constructed with the following components. They are:

1. **Instructions/Statements:** This is said to be known as the heart of an algorithm, which contains a sequence of sub-algorithms that could perform minor task. Every algorithm could be separated in the form of instructions that is similar to an object which is composed of atoms. Certain algorithms could not be separated in the form of instructions, say digit addition, which could form as a fundamental instruction.

   Statements contain the following contents:

   i. **Series:** All the Steps should be done in particular order, and each of the steps must be used.
   ii. **Specific:** Any Step must not he replaced by similar step.
   iii. **Steps:** Everything we do is called step.
   iv. **Solve:** Write an algorithm to get output from the input.

   Simple example:

   Multiplication of two numbers:

   a. **Description of Problem:** To find multiplication of two numbers.
   b. **Setup:** Two numbers required for multiplication for storing the results.
   c. **Parameters:** I. Read first number 2. Read second number.
   d. **Execution:** result = a*b.
   e. **Conclusion:** The desired output is result.

2. **Sequence:** A good algorithm should be written in a sequenced manner, where each step can be executed exactly one time.

An algorithm to solve the problem of converting Fahrenheit to Celsius is:

- Read the temperature value in Fahrenheit.
- Calculate the Celsius using the formula:

Celsius= (5/9) * (Fahrenheit-32)

- Display result in degree Celsius.

3. **Variables:** These provide a mean to name values so that they can be used and manipulated later on.

   >>English = 56

   Here variables English refers to the value 56

4. **Selection/Conditional:** Algorithms could select a particular instruction based on certain condition. In selection statements, only one of the statements are executed based on the condition.

   There are three kinds of control flow structures:

   i. **Sequence:** These refers to one or more instructions that the computer performs in sequential order.

      Example: Algorithm:

      - **Step 1:** Start
      - **Step 2:** Read the two values to the variables first and second
      - **Step 3:** Calculate the addition

        Sum = first + second
      - **Step 4:** Print the result Sum
      - **Step 5:** Stop

      Pseudocode:

      READ first, second
      CALCULATE sum =first + second
      print(sum)

   Flowchart:

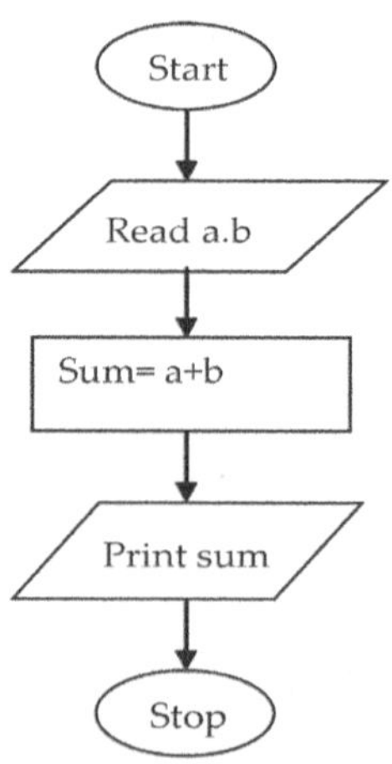

Program:

```
first = 5
second = 4
add = first + second
```

print("The sum of two numbers is:,"add)

ii. **Selection:** It is making a decision among several actions based on some conditions.

Algorithm:

- **Step 1:** Start
- **Step 2:** Get the value of a
- **Step 3:** Check whether the value of a is equal to zero
- **Step 4:** Print the result as negative
- **Step 5:** Stop

Pseudocode:

```
READ a
IF a == 0 THEN
        print("a is negative")
END IF
```

Flowchart:

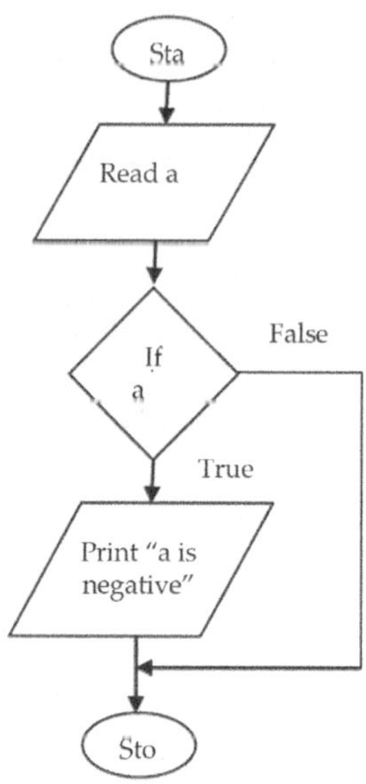

Example:

```
a=0
if a==0:
        print("Negative Number")
```

iii. **Iteration:** A Loop is one or more instructions that the computer performs repeatedly (repetition or loop).

Algorithm:

- **Step 1:** Start
- **Step 2:** Get the value for limit
- **Step 3:** Test out whether the limit reached
- **Step 4:** Print "Good morning"
- **Step 5:** Go over the Steps from 3 to 4
- **Step 6:** Stop

Pseudocode:

```
READ limit
FOR i IN RANGE(0,limit)
        Print("GOOD NIGHT")
END
```

Flowchart:

Example:

```
for I in range(0,5):
        print("GOOD NIGHT")
```

Output:

```
GOOD NIGHT
GOOD NIGHT
GOOD NIGHT
GOOD NIGHT
GOOD NIGHT
```

5. **State:** It is the particular condition that a variable is in at a specific time.
6. **Repetition/Control Flow:** One or more steps is executed again and again for the specified number of times

   Example: Find average of first five numbers using for loop

```
total=0
for i in range I to 5
        total= total+i
end for
average = total / 5
print average
```

7. Functions/Sub-Algorithms:
    - It is a collection of statements that is used to carry out some task. It is easy to write function and easy to find out error in the program if we are using function.
    - In many cases algorithms will not perform its task in its own. Those complicated algorithms could be separated into smaller ones, which could be very easy to refer or use further.
    - For example, in the algorithm of finding the prime numbers between two integers, the prime number checking function could be reused. Breaking up of algorithm into logical parts could be very easy to analyze their behavior and properties from a mathematical point of view
8. **Performance:** This analysis helps us to select the best algorithm from many algorithms to solve a problem. To compare algorithms, a set of parameters are considered. Generally, the performance of an algorithm depends on the following aspects:
    - Whether the algorithm is providing the exact solution for the problem?
    - Whether it is easy to understand and implement?
    - How much space (memory) it requires to solve the problem?
    - Too much time it takes to solve the problem?

Performance analysis of an algorithm is computed by using the following measure:

- o Space required for completing the task (Space Complexity). It includes program space and data space
- o Time required completing the task of that algorithm (Time Complexity)

3.9. NOTATION (PSEUDOCODE, FLOW CHART, PROGRAMMING LANGUAGE)

3.9.1. Pseudocode

Pseudo consists of short legible and properly styled English language used for amplification an algorithm. It uses plain English statements rather than symbols. It cannot be compiled and executed. It is also called "program design language [PDL].

Example:

A pseudocode to add two numbers and display the results:

```
READ num1, num2
result = num1 + num2
WRITE result.
```

Rules for writing pseudocode:

- Only one statement can be written per line
- Initial keywords must be in Capital letters (READ, WRITE, IF, WHILE, UNTIL).
- Pseudocode should be aligned properly to show hierarchy.
- Multiline structures must be ended at the last

➢ **Advantages:**

- It can be done effortlessly in any word processor;
- It is very easy to create and modify when compared to the flow chart;
- It is user-readable and very easy to understand;
- It is very undemanding to translate a pseudocode to a programming language.

➢ **Disadvantage:**

- It is not visually illustrated;
- This does not form a pictorial form of representation with an exact style and format;
- This method could be very complicated for the beginners of programming language;

3.9.2. Basic Guidelines for Writing Pseudocode:

1. Write Only One Statement per Line: The pseudocode should contain the statement that it should show only one action to the computer. Each and every task should correspond to each line of code.

2. Capitalize Initial Keyword: The keywords should be written in capital letters. For example, READ, and WRITE. Following are certain keywords that are commonly used: WHILE; ENDWHILE; REPEAT; UNTIL; IF; ELSE; ENDIF.

3. Indent to Show Hierarchy: In our design structure, it is necessary to follow certain indentation pattern. Indentation is a process of showing the boundaries of the structure:

- *Sequence:* It keeps the statements, which are "stacked" in a sequence that starts in a similar column.
- *Selection:* The indentation is carried out in the statement and not in the keywords of the selected structure.
- *iLooping:* The indentation is conceded for the statements that comes inside the loop, but not the keywords that form the loop.

4. End Multiline Structures: Each structure must be ended properly, which provides more clarity. *Example:* ENDIF for IF statement

5. Keep Statements Language Independent: There may be special features available in the language that you plan to eventually write the program in. The code should be written in the most convenient language of the user. *Examples:* Pseudocode:

- **Problem 1:** Calculate sum and average for n numbers.

```
BEGIN
INITIALIZE add=0, i=1
READ n
FOR i <=n, then
        COMPUTE add = add +i
        CALCULATE i=i+1
END FOR
COMPUTE avg = add/n
PRINT add, avg
END
```

- **Problem 2:** Calculate area of circle:

```
BEGIN
READ radius, r
INITIALIZE pi=3.14
CALCULATE Area=pi * r *r
PRINT Area
END
```

- **Problem 3:** Read Number n and print the integers counting up to n

```
BEGIN
READ n
INITIALIZE i to 1
FOR i <= n, then
```

```
        DISPLAY i
        INCREMENT i
END FOR
END
```

- **Problem 4:** Find the greater number between two numbers.

```
BEGIN
Read a, b
IF a is less than b THEN
        BIG = b
        SMALL =a
ELSE
        BIG = a
        SMALL = b
WRITE / DISPLAY "BIG, SMALL"
END IF
```

- **Problem 5:** To determine a student whether successful or fail.

```
BEGIN
READ student grade
IF student's grade is greater than or equal to 50 THEN
        Print "passed"
ELSE
        Print "failed"
END IF
```

3.9.3. Flowchart-Introduction

A flow chart is a diagrammatic demonstration of an algorithm. The benefits of flowcharts are as follows:

- It is easy to understand;
- A problem can be analyzed easily with flowchart;
- It gives clear idea of a program;
- It acts as a guide during the development of the program;
- It helps to clear the errors in coding;
- It helps in maintenance of code.

Restrictions of using flowcharts:

1. **Complex Logic:** The flow chart becomes more complicated when the program logic is complex.
2. **Alterations and Modifications:** Certain changes done in the program will make the entire flowchart to collapse.
3. **Reproduction:** Since the symbols of the flowchart cannot be converted to be executed, imitation of flowchart becomes a trouble.
4. The requirement of what is done could be gone astray effortlessly in the technical information of how it is done.

3.9.3.1. Flowcharts Symbols

Certain symbols are used in the graphical representation of the flowchart, which are as follows:

1.	Terminal symbols		Represent the start and stop of the program.
2.	Input / Output		Denoted either an input or output operation.
3.	Process symbol	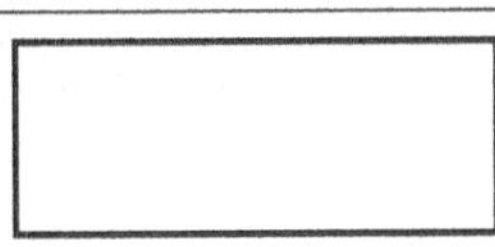	Denotes the process to be carried out.
4.	Decision	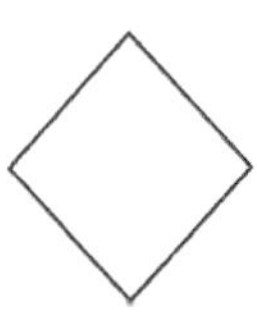	Represent decision making and branching.
5.	Flow lines	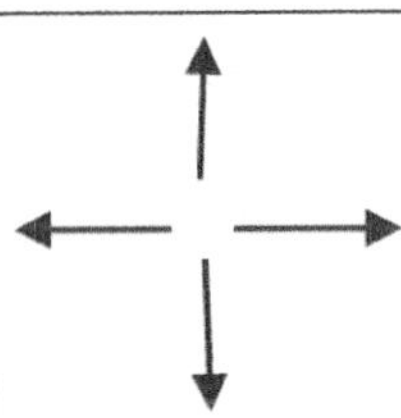	Represents the progression of steps and way of flow. Used to attach symbols.

6.	Connectors		A connector symbol is a circle with the text inside to identify the link. This symbol are used to connect the flowchart.

3.9.3.2. Basic Guide Lines for Preparing Flowchart

- A particular logical order should be maintained and listed out while drawing a flow chart.
- The flowchart should be understandable, neat, and concise. It should not lead to ambiguity.
- The direction of the flowchart is from top towards bottom or left toward right.
- It is necessary to have a single flow line that enters out from the process symbol.
- At the same time, a single incoming flow line alone could enter the decision symbol. But it can have two or three flow lines coming out from the symbol.

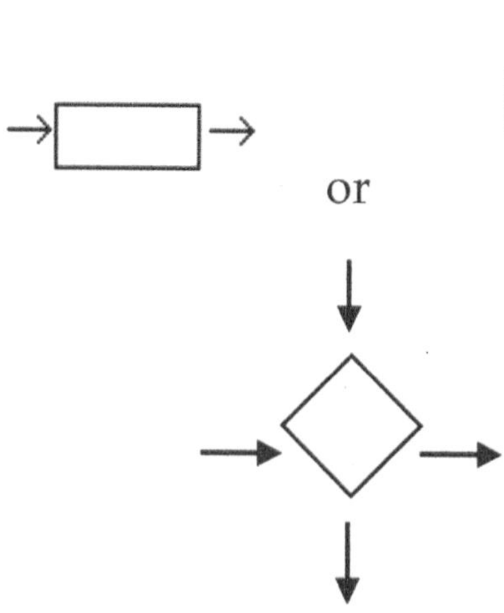

- Terminal symbol should have a single flow line.

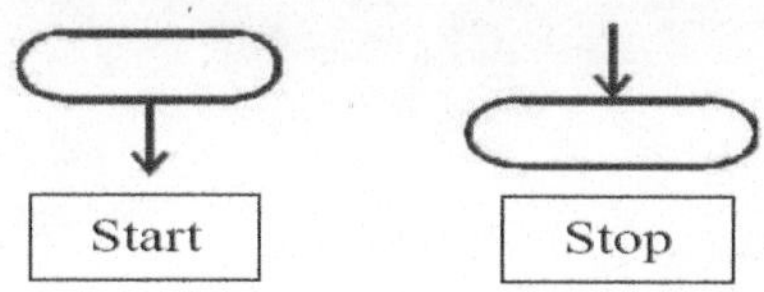

- Explanation can be written within the symbol. If required, use the annotation symbol to portray data or computational steps obviously.

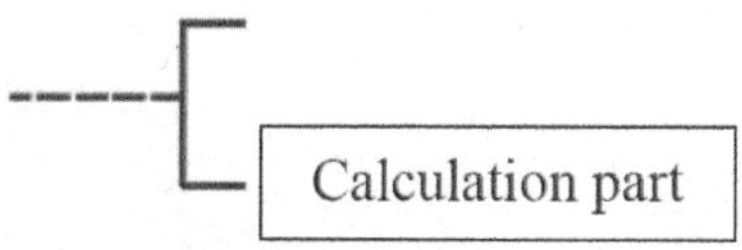

- Connector symbols could be used when the flowchart becomes intricate to decrease the flow lines. For a better communication, it is necessary to evade the intersection of flow lines.
- The flowchart must have a logical start and stop.
- By passing from end to end it with a simple test data, the validity of the flowchart can checked out

3.9.3.3. Example for Algorithm, Pseudocode, Flowchart

1. Draw the flowchart to find the largest among A and B.

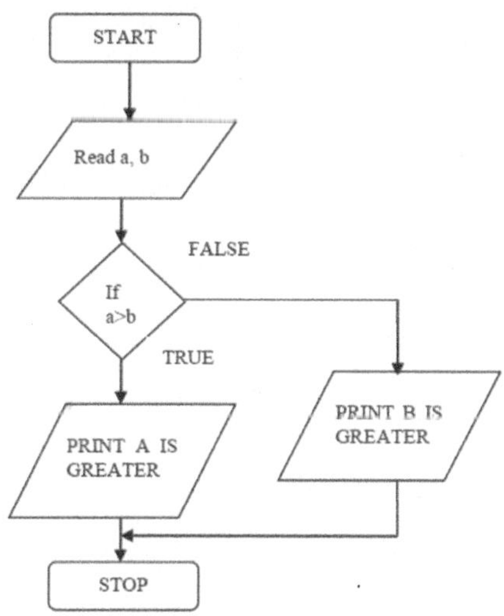

Algorithm:

- **Step 1:** Start
- **Step 2:** Read a, b, c
- **Step 3:** Compare the values of a, b
- **Step 4:** If a is greater than b, then display "a is greater"
- **Step 5:** Otherwise display "b is greater"
- **Step 6:** Stop

Pseudocode Read a, b

```
IF a > b THEN
        WRITE a is big
ELSE
        WRITE b is big
ENDIF
```

2. Find the area of a circle of radius r.

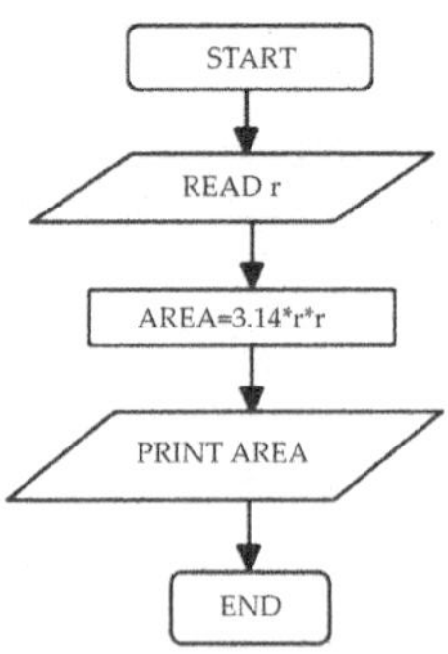

3. Convert temperature Fahrenheit to Celsius.

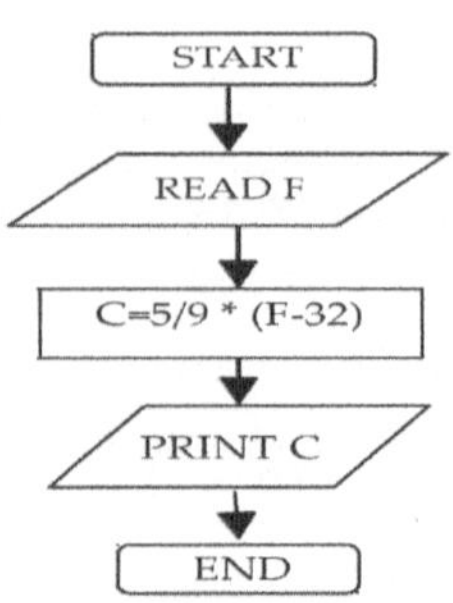

4. Flowchart for an algorithm which gets two numbers and prints the sum of their value.

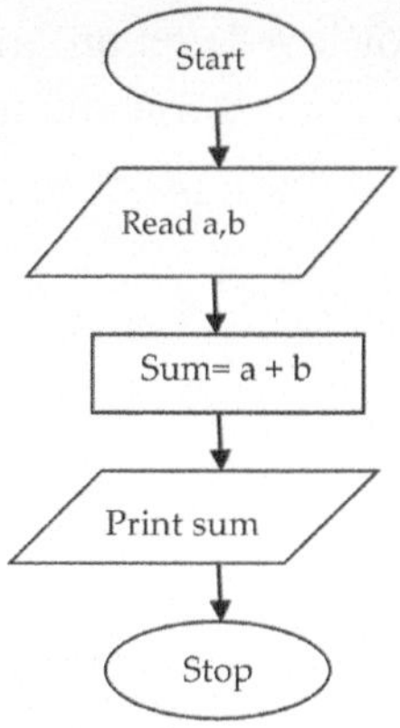

3.9.4. Programming Language

A computer is electronic and supreme machine to perform computational algorithms since:

- Arithmetic operations can be performed by the computer.
- The operations could be performed only if certain conditions are fulfilled (with the help of conditional branch instruction).

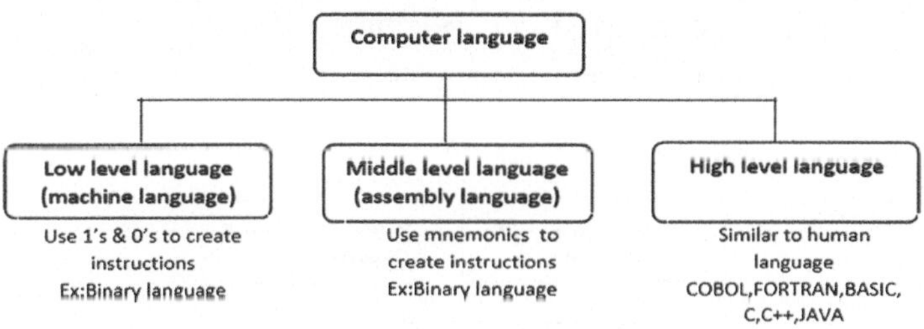

Classification of languages used in computer programming:

1. Machine Language:
   a. It is the binary number (0,1) which is converted from the simple language as machine-understandable.
   b. All Central Processing Unit has its own machine understandable language.
   c. i.e., the instruction code 1 can convert a dissimilar instruction for different CPUs.
   d. What a machine language program looks like:
      11110011
      10100111
      11111111
      10010100

Assembler language or low-level programming language:

- An assembly language is a mnemonic which is like an English
- Every machine instruction is an example for mnemonics

Example:

start
add
sub
....
....
x, y
x, y

Every assembly instruction is a mnemonic that corresponds to an exclusive machine instruction

2. High Level Programming Language

A high level language is the one which can be understandable by the human which is used to

This allows programmer to write the code in his language that could be effortlessly converted into machine instructions

iStatements are the sentences which are written in a high level language.

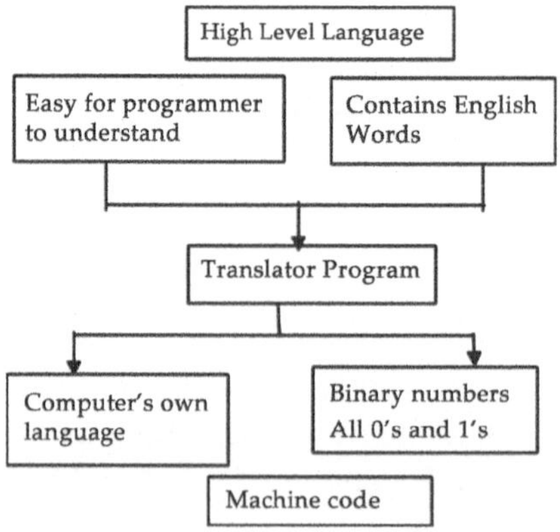

Example Program:

```
Void main()
    {
        if (x1 > y1)
    {
        maximum = x1;
```

```
    }
    else
    {
            maximum = y1;
    }
    .....
    }
```

Some well-known programming languages

Name	Application Area
Fortran	Scientific application (FORmula TRANslator)
Cobol	Business application (common business-oriented language)
C	System application (successor of a language called "B")
C++	System application (successor of the language "C")
Java	General-purpose (a generalization of "C++")
C#	All-purpose purpose (another simplification of "C++")
Perl	Scripting language with many string processing potential
Python	Script language with indentation as block denotation.

3.10. ALGORITHMIC PROBLEM SOLVING

Algorithm forms as a technical solution to a problem. These solutions are not clarification but specific instructions for receiving answers.

1. Understanding the Problem: The very first step for the problem solving is the understanding of the given problem.

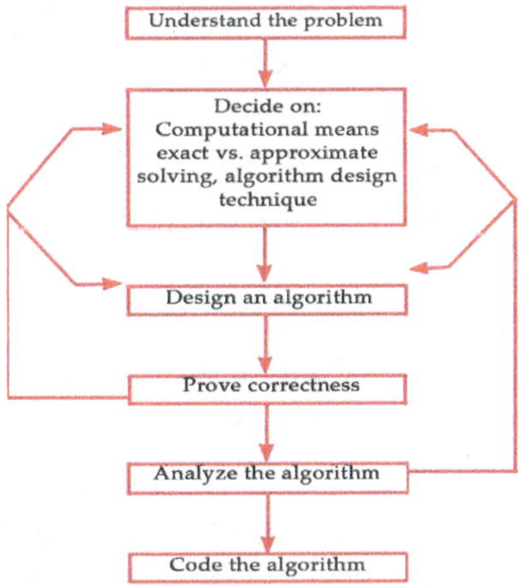

Go through the problem's narrative carefully and raise queries in case of any doubt arises, perform a few small examples by hand, consider about special cases, and ask questions over again if essential.

An input to an algorithm specifies an occurrence of the problem the algorithm solves. It is important to identify exactly the range of instances the algorithm needs to hold.

2. Determine the Potential of the Computational Device: If you have completely understood the given problem, you have to determine the potential of the device which is proposed for.

- **Sequential Algorithms:** The instructions are carried out one by one at a time. Similarly, an algorithm is designed in a certain manner to be executed on a particular machine.
- **Parallel Algorithms:** The central supposition of the RAM model does not cling to some novice computers that can execute operations concomitantly, i.e., in corresponding.

3. Preferring Between Exact and Approximate Problem Solving: The next principal decision is to select how to solve the problem:

- If the problem is solved accurately is called an exact algorithm;
- If the problem is solved, something like is called an approximation algorithm.

Need of approximation algorithm:

- The first is some of the specific problems which could not be solved simply based on definite instances; examples embrace finding square roots, explaining nonlinear equations, and appraising definite integrals.
- Second, because of the problem complication, the available algorithms for that particular problem could be very deliberate. This problem takes place when the problem has a vast amount of choices.
- Third, an approximation algorithm could be an appropriate penchant for solving the most difficult problem.

4. Deciding on Appropriate Data Structures: In the pioneering world of object-oriented programming, data structures become unavoidably significant for both design and analysis of algorithms. Data structures helps in organizing and storing the data

Algorithms + Data Structures = Programs

5. Algorithm Design Techniques: It is a universal right to use to resolve problems algorithmically, which is germane to a variety of problems from various area of computing. A most important one is that they afford assistance for scheming algorithms to new problems, i.e., the problems, which has unsatisfactory algorithm. Next, algorithms are the foundation stone of computer science. Every science is alarmed in categorizing its chief subject, and computer science has no exemption. Algorithm design techniques construct it potential to categorize algorithms based on a fundamental design idea; therefore, they can provide a normal way to both categorize and study algorithms.

6. Methods of Denoting an Algorithm: Once an algorithm is designed, you need to denote it in some fashion.

i. Complimentary and also a step-by-step form;
ii. Pseudocode.

There are two methods tailored recently to specify algorithms.

Pseudocode is a combination of a normal language and programming language like constructs. Pseudocode is normally more explicit than natural language, and its procedure regularly yields to more concise algorithm similes.

Previously flowcharts were used as a leading procedure for the specification of algorithms. This flowchart uses certain geometric shapes for expressing the algorithm's procedure. For simple algorithms, this method was most adaptable but for the complex one, this was not very sufficient.

7. Demonstrating an Algorithm's Accuracy: Formerly an algorithm has been precise; it has to be confirmed for its correctness. That is, you must verify that the algorithm yields a mandatory result for every genuine input in a limited amount of time.

A frequent technique for demonstrating the correctness involves the usage of mathematical induction since an algorithm's iterations provide a natural sequence of steps necessary for certain proofs. Although tracing the algorithm's recital for certain specific inputs can be a very worthwhile activity, it cannot confirm the algorithm's correctness realistically.

The concept of precision for approximation algorithms is less uncomplicated than it is for exact algorithms since the error produced in this approximation algorithm should not expand the firm limit.

8. Analyzing an Algorithm: Efficiency is a significant characteristic of any algorithm.

There are two variety of algorithm competence:

- **Time Efficiency:** It shows how fast the algorithm runs
- **Space Efficiency:** It shows how much extra memory it uses.

Another desirable characteristic of an algorithm is simplicity.

Because simpler algorithms are uncomplicated to understand and easier to program; accordingly, the resulting programs frequently have fewer bugs. Sometimes simpler algorithms are also more resourceful than more complicated alternatives.

Simple strategies for developing algorithms (iteration, recursion)

There are two fundamental strategies to solve a problem:

- Top-down approach (recursion); and
- Bottom-up approach (iteration).

In simple words, Iteration, in the framework of computer programming, is a process wherein a set of instructions or structures are recurring in a sequence a specified number of times or until a condition is met. When the first set of instructions is executed again, it is called iteration.

Recursion is a problem solving approach by which a function calls itself repeatedly until some specified condition has been satisfied. Recursion splits a problem into one or more simpler versions of itself.

Recursion	Iteration
Repetition is achieved through repeated function calls	Iteration is explicitly a repetition structure
Recursion expires when a base case is recognized	Iteration terminates when the loop continuation test turns out to be false
Recursion causes another copy of the function and hence a considerable memory space is occupied	Iteration usually occurs surrounded by a loop, so the extra memory assignment is omitted

3.11. FLOW OF CONTROL

Three basic constructs for flow of control:

1. **Sequence:** It is a default mode. It is used for the sequential execution of statements, one line after another.
2. **Selection:** It is Used for decisions to choose between two or more alternative paths.
3. **Repetition:** It is Used for looping to repeat a piece of code several times.

3.11.1. Sequence

It is specified by writing one statement after another, each statement on a line by itself, and all statements aligned with the same indent. The actions are executed in the order in which they are written, from top to bottom.

➢ **Common Keywords:**

- Input: READ, INPUT, OBTAIN, GET
- Output: PRINT, OUTPUT, DISPLAY, SHOW
- Compute: COMPUTE, CALCULATE, DETERMINE
- Initialize: SET, INIT

Add one: INCREMENT, BUMP

Examples:

1. Pseudocode for computing the area of a rectangle.

```
READ rectangle height
READ rectangle width
COMPUTE area =height* width
PRINT area
```

2. Pseudocode for computing the average of five numbers.

```
PRINT "Enter the 5 numbers"
READ m1, m2, m3, m4, m5
PRINT "The average is"
SET average to (m1+m2+m3+m4+m5)/5
        PRINT average
```

3.11.2. Selection

When the choice is made between two alternative courses of action it is called decision (selection) and it comprises of the following constructs:

i. IF-THEN-ELSE
ii. CASE...ENDCASE

➢ **IF-THEN-ELSE**

Binary choice is indicated by the use of four keywords:

IF, THEN, ELSE, and ENDIF.

The general form is:

```
IF condition THEN
        sequence 1
ELSE
        sequence 2
ENDIF
```

The ELSE keyword and "sequence 2" are optional. If the condition is true, sequence 1 is performed, otherwise sequence 2 is performed.

Examples:

i. Pseudocode to check whether the number is odd or even.

```
READ number
IF number MOD 2 = 0 THEN
DISPLAY "Number is Even"
ELSE
DISPLAY "Number is Odd"
ENDIF
```

ii. Pseudocode to check whether the given non-zero number is positive or negative.

```
READ number
IF num is less than 0 THEN
PRINT num is negative
ELSE
PRINT num is positive
ENDIF
```

CASE

CASE is a multiway branch (decision) based on the value of an expression. CASE is a simplification of IF-THEN-ELSE. Four keywords, CASE, OF, OTHERS, and ENDCASE, and conditions are used to point to the various alternatives.

The general form is:

```
CASE expression OF
        condition 1: sequence 1
        condition 2: sequence 2
        ...
```

condition n: sequence n
OTHERS:
default sequence
ENDCASE

The other clause with its default sequence is non-compulsory. Conditions are usually numbers or characters indicating the value of "expression," but they can be English statements or some other notation that indicates the condition under which the given sequence is to be performed. A certain sequence may be associated with more than one condition.

Examples:

iii. Pseudocode for simple calculator

```
READ m1, m2
READ choice
CASE choice OF
+: PRINT m1+m2
–: PRINT m1-m2
*: PRINT m1*m2
/: PRINT m1/m2
ENDCASE
```

iv. Pseudocode for determining grade points from grades.

```
READ grade
CASE grade OF
S: gradepoint = 10
A: gradepoint = 9
B: gradepoint = 8
C: gradepoint = 7
D: gradepoint = 6
E: gradepoint = 5
U: gradepoint = 0
ENDCASE
DISPLAY gradepoint
```

3.11.3. Repetition

It is a loop (iteration) based on the satisfaction of some condition(s). It comprises of the following constructs:

- WHILE...ENDWHILE
- REPEAT...UNTIL
- FOR...ENDFOR

- **WHILE...ENDWHILE**

The body of the statements will be executed again and again until some conditions are satisfied. The start and end of the loop are specified by two keywords WHILE and ENDWHILE.

The general form is:

```
WHILE condition
        sequence
ENDWHILE
```

The loop is enters into the body only if the condition is true. The "sequence" is performed for each iteration. At the conclusion of each iteration, the condition is appraised, and the loop continues as long as the condition is true.

Examples:

i. Pseudocode to print the numbers from 1 to 100.

```
n=1
WHILE n is <= to 100
        DISPLAY n
        INCREMENT n by 1
ENDWHILE
```

ii. Pseudocode to print the sum of the digits of a given number

```
INPUT number
INITIALIZE Sum as zero
WHILE Number is not zero
        COMPUTE Remainder by Number Mod 10
        ADD Remainder to Sum
        DIVIDE Number by 10
ENDWHILE
PRINT Sum
```

- **REPEAT...UNTIL**

This is similar to WHILE except that the condition is tested at the end of the loop. It have two keywords, REPEAT, and UNTIL.

The general form is:

```
REPEAT
        statement
UNTIL condition
```

The "statement" in this type of loop is always performed at least once, because the test is carried out after the sequence is executed. At the conclusion of each iteration, the condition is estimated, and the loop repeats if the condition is false. The loop terminates when the condition becomes false.

Examples:

i. Pseudocode to print the numbers from 1 to 100.

```
n=1
REPEAT
        DISPLAY n
        INCREMENT n by 1
UNTIL n is greater than 100
```

ii. Pseudocode to print the sum of the digits of a given number

```
INPUT a Number
INITIALIZE Sum to zero
REPEAT
        COMPUTE Remainder = Number % 10
        ADD Remainder to Sum, sum=sum+Remainder
        DIVIDE Number by 10, Number=Number/10
UNTIL Number = 0
PRINT Sum
```

➢ **FOR...ENDFOR**

It is a "counting" loop. This loop is a unique construct for iterating a specific number of times, often called a "counting" loop. Two keywords, FOR, and ENDFOR are used.

The general form is:

```
FOR iteration condition
        statements
ENDFOR
```

Examples:

i. Pseudocode to print the numbers from 1 to 100.

```
FOR n=1 to 100
        DISPLAY n
ENDFOR
```

ii. Pseudocode to input ten numbers and print the sum.

```
INITIALIZE sum to 0
FOR n=1 to 10
        INPUT number
        COMPUTE sum as sum+number
ENDFOR
DISPLAY sum
```

3.12. ILLUSTRATIVE PROGRAM

1. Guess an Integer in a Range

Algorithm:

- **Step1:** Start.
- **Step 2:** Declare hidden, guess.
- **Step 3:** Compute hidden= Choose a random value in a range.
- **Step 4:** Read guess.
- **Step 5:** If guess = hidden, then
  Print Guess is hit
  Else
  Print Guess not hit Print hidden
- **Step 6:** Stop

Pseudocode:

```
BEGIN
COMPUTE hidden = random value in a range
READ guess
IF guess = hidden, then
    PRINT Guess is hit
ELSE
    PRINT Guess not hit
    PRINT hidden
END IF-ELSE
END
```

Flowchart:

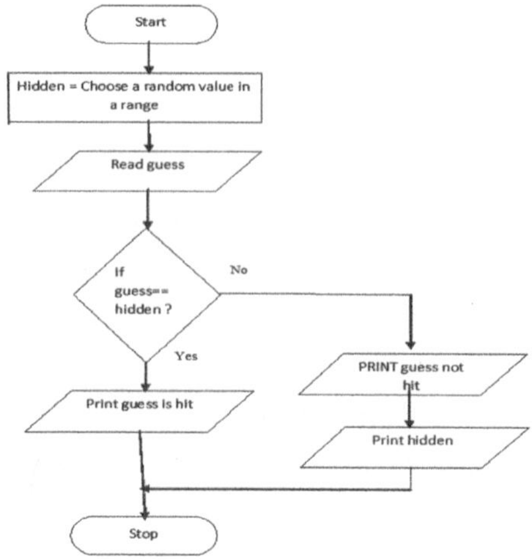

2. Find Minimum in a List

Algorithm:

- **Step 1:** Start
- **Step 2:** Read the limit n
- **Step 3:** Initialize I to 0
- **Step 4:** If the values of i<n, then go to the step 4.1,

else go to step

- o **Step 4.1:** Read a[i]
- o **Step 4.2:** i=i+1 go to the step 4
- **Step 5:** Compute min=a[0]
- **Step 6:** Initialize the value i=1
- **Step 7:** If i<n, then go to step 8 else go to step 10
- **Step 8:** If a[i]<min, then go to step 8.1,8.2 else go to 8.2
- o **Step 8.1:** min=a[i]
- o **Step 8.2:** i = i + 1 go to 7
- **Step 9:** Print min
- **Step 10:** Stop

Pseudocode:

```
BEGIN
READ n
FOR i=0 to n, then
        READ a[i]
        INCREMENT i
END FOR COMPUTE
min=a[0]
FOR i=1 to n, then
        IF a[i]<min, then
                CALCULATE min=a[i]
                INCREMENT i
        ELSE
                INCREMENT i
END IF-ELSE END FOR
PRINT min
END
```

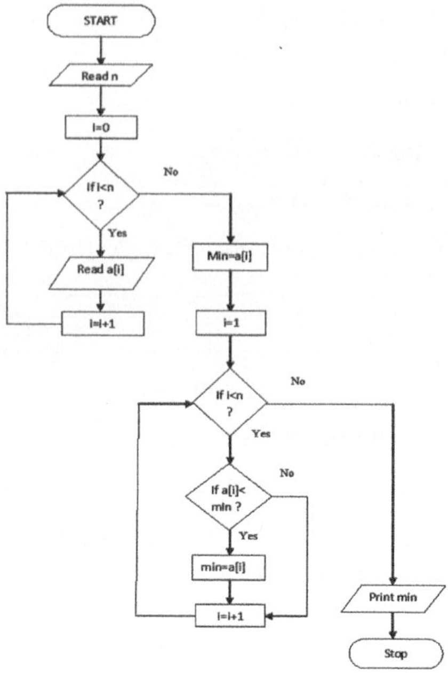

3. Insert a Card in a List of Sorted Cards

Algorithm:

- **Step 1:** Start
- **Step 2:** Read n
- **Step 3:** Initialize i=0
- **Step 4:** If i<n, then go to step 4.1, 4.2 else go to step 5
    - o **Step 4.1:** Read a[i]
    - o **Step 4.2:** i=i+1 go to step 4
- **Step 5:** Read item
- **Step 6:** Calculate i=n–1
- **Step 7:** If i>=0 and item<a[i], then go to step 7.1, 7.2 else go to step 8
    - o **Step 7.1:** a[i+1]=a[i]
    - o **Step 7.2:** i=i–1 go to step 7
- **Step 8:** Compute a[i+1]=item
- **Step 9:** Compute n=n+1
- **Step 10:** If i<n, then go to step 10.1, 10.2 else go to step 11
    - o **Step 10.1:** Print a[i]
    - o **Step 10.2:** i=i+1 go to step 10

- **Step 11:** Stop

Pseudocode:

```
BEGIN
READ n
FOR i=0 to n, then
        READ a[i]
        INCREMENT i
END FOR READ item
FOR i=n–1 to 0 and item<a[i], then
        CALCULATE a[i+1]=a[i]
        DECREMENT i
END FOR COMPUTE
a[i+1]=a[i]
COMPUTE n=n+1
FOR i=0 to n, then
        PRINT a[i]
        INCREMENT i
END FOR END
```

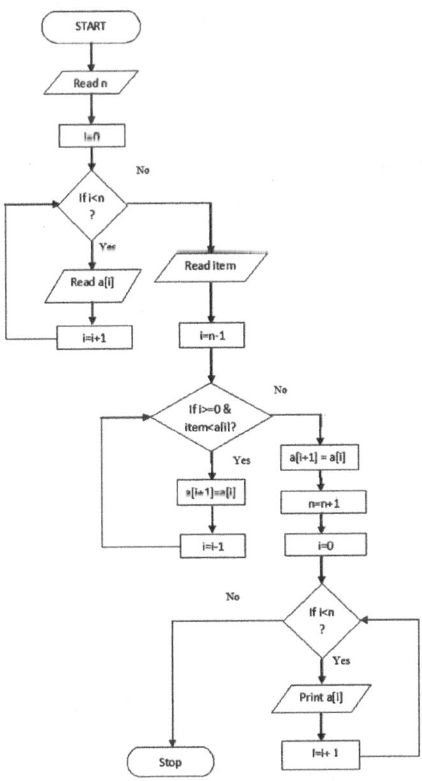

4. Tower of Hanoi:

Algorithm:

- **Step 1:** Start
- **Step 2:** Read n
- **Step 3:** Calculate move=pow(2,n)–1
- **Step 4:** Function call T(n,Beg,Aux,End) recursively until n=0
  - o **Step 4.1:** If n=0, then go to step 5 else go to step 4.2
  - o **Step 4.2:** T(n–1,Beg,End,Aux) T(1,Beg,Aux,End), Move disk from source to destination T(n–1,Aux,Beg,End)
- **Step 5:** Stop

Pseudocode:

```
BEGIN READ n
CALCULATE move=pow(2,n)–1
FUNCTION T(n,Beg,Aux,End) Recursively until n=0
PROCEDURE IF n=0 then,
        No disk to move
Else
        T(n–1,Beg,End,Aux)
        T(1,Beg,Aux,End), move disk from source to destination
        T(n–1,Aux,Beg,End)
END PROCEDURE
END
```

Flowchart:

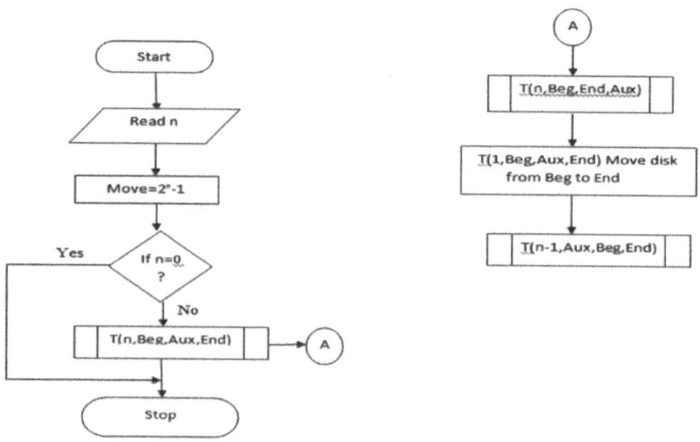

3.12.1. Procedure to Solve Tower of Hanoi

The goal of the puzzle is to move all the disks from leftmost peg to rightmost peg:

- Move only one disk at a time;
- A larger disk may not be placed on top of a smaller disk. For example, consider n=3 disks.

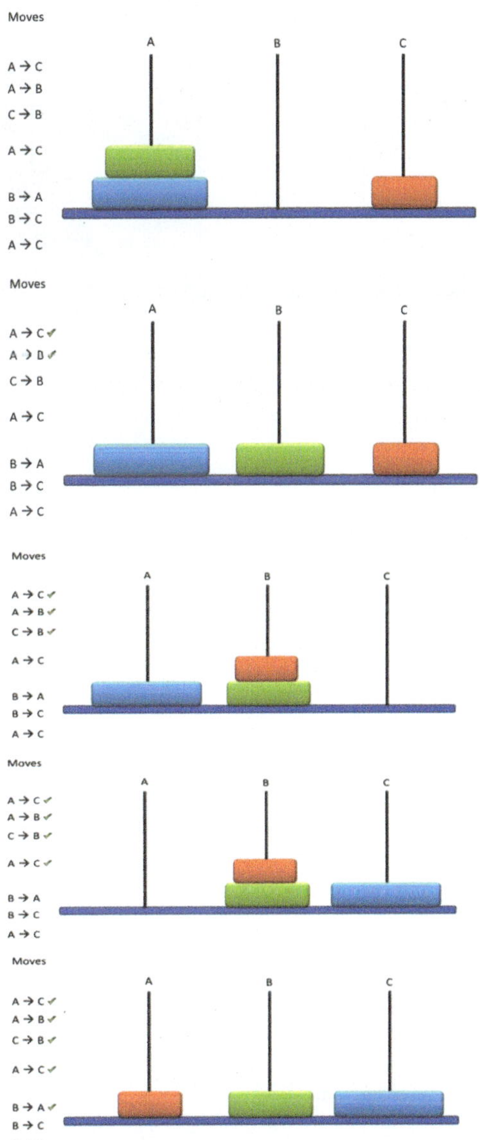

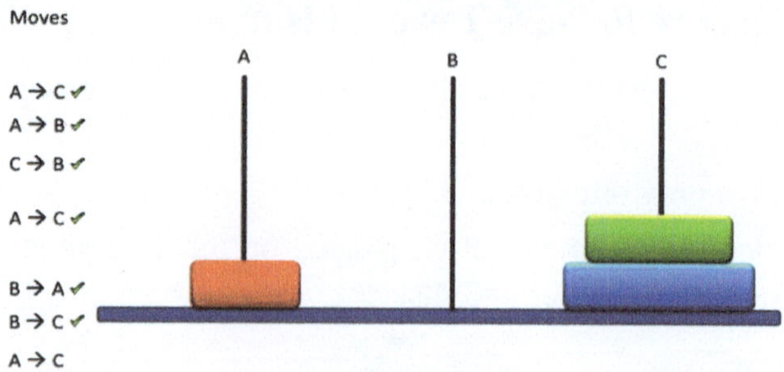

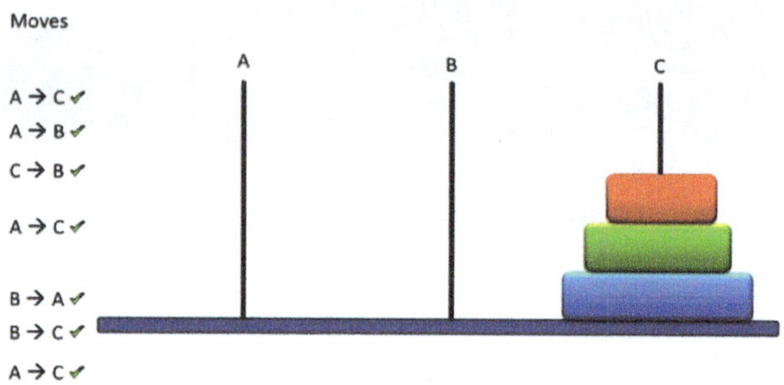

Data, Expressions, and Statements

CHAPTER 4

PYTHON PROGRAMMING: AN INTRODUCTION

CONTENTS

4.1. INTRODUCTION TO PYTHON

4.1.1. What Is Python?

Python is a general-purpose interpreted, interactive, object-oriented, and high-level programming language. It was founded by Guido van Rossum during 1985–1990. Python got its name from "Monty Python's flying circus." Python was released in the year 2000.

4.1.2. Features of Python

1. **Python is Interpreted:** Python is managed at runtime by the interpreter.
2. **Python is Interactive:** You can interact with the interpreter straightly to write your programs.
3. **Python is Object-Oriented:** Python supports Object-Oriented style or technique of programming that encapsulates code within objects.
4. **Python is a Beginner's Language:** Python is mainly for the beginner:
    i. Level programmers and supports the improvement of an extensive variety of applications.
5. **Easy-to-Learn:** Python is easily readable. The structure of the program is very simple. It employs few keywords.
6. **Easy-to-Maintain:** Python's source code is fairly easy-to-maintain.
7. **Portable:** Python can run on a wide variety of hardware platforms and has the same interface on all platforms.
8. **Interpreted:** Python is processed at runtime by the interpreter. So, there is no need to compile a program before executing it. You can simply run the program.
9. **Extensible:** Programmers can embed python within their C,C++,Javascript, ActiveX, etc.
10. **Free and Open Source:** Anyone can freely distribute it, read the source code, and edit it.
11. **High Level Language:** When writing programs, programmers concentrate on solutions to the current problem, no need to worry about the low level details.
12. **Scalable:** Python provides a better structure and support for large programs than shell scripting.

4.1.3. History of Python

The history of Python commences with ABC. ABC is a general-purpose programming language and programming environment, which had been developed in the Netherlands, Amsterdam, at the CWI (Centrum Wiskunde and Informatica). The ultimate accomplishment of ABC was to inspire the design of Python.

Python was conceptualized in the late 1980s. Guido van Rossum worked that time in a project at the CWI, called Amoeba, a distributed operating system. He programmed in ABC. In an interview with Bill Venners (January 2003), Guido van Rossum said: "I remembered all my experience and some of my frustration with ABC. I decided to try to design a simple scripting language that possessed some of ABC's better properties, but without its problems. So I started typing. I created a simple virtual machine, a simple parser, and a simple runtime. I made my own version of the various ABC parts that I liked. I created a basic syntax, used indentation for statement grouping instead of curly braces or begin-end blocks, and developed a small number of powerful data types: a hash table (or dictionary, as we call it), a list, strings, and numbers.

4.1.4. Comparing Python with Java, Perl, and Other Programming Languages

Prof. Lutz Prechelt from the University of Karlsruhe compared Python with other programming languages. He sums up his results: "80 implementations of the same set of requirements are compared for several properties, such as run time, memory consumption, source text length, comment density, program structure, reliability, and the amount of effort required for writing them. The results indicate that, for the given programming problem, which regards string manipulation and search in a dictionary, 'scripting languages'

(Perl, Python, Rexx, Tcl) are more productive than 'conventional languages' (C, C++, Java). In terms of run time and memory consumption, they often turn out better than Java and not much worse than C or C++.

4.1.5. Other Advantages of Python

It's amazingly calm to embed Python, or better the Python interpreter into C programs. By doing this, you can add features from Python that could take months to code in C. Vice versa, it's likely to spread the Python interpreter by including a module written in C. One reason to do this is if a C library exists that does something which Python doesn't. Another good reason is if you need something to run quicker than you can achieve in Python.

The Python Standard Library comprises of huge number of valuable elements and is part of every standard Python installation. After having learned the fundamentals of Python, it is essential to become aware of the Python Standard Library because many problems can be resolved rapidly and effortlessly if you are familiar with the potentials that these libraries provide.

4.2. DOWNLOADING AND INSTALLING PYTHON 3.6.2

Installing Python is simple, and today numerous Linux and UNIX distributions comprises of fresh Python. There are inbuilt Python in some Windows.

4.2.1. Python: Version 3.6.2

The Python download needs about 30 Mb of disk space; keep it on your machine, in case you want to re-install Python. When installed, Python needs an extra 90 Mb of disk space.

4.2.2. Python on Windows

Downloading:

- Go to https://www.python.org/downloads/.

  See below figure:

Click the Download Python 3.6.2 button.

The file named python-3.6.2.exe should start downloading into your standard download folder.

See below figure:

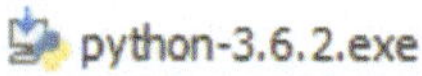

- Move this file to a more permanent location to install Python
- Start the Installing instructions directly below.

Installing:

- Double-click the icon labeling the file python-3.6.2.exe.

  See below figure:

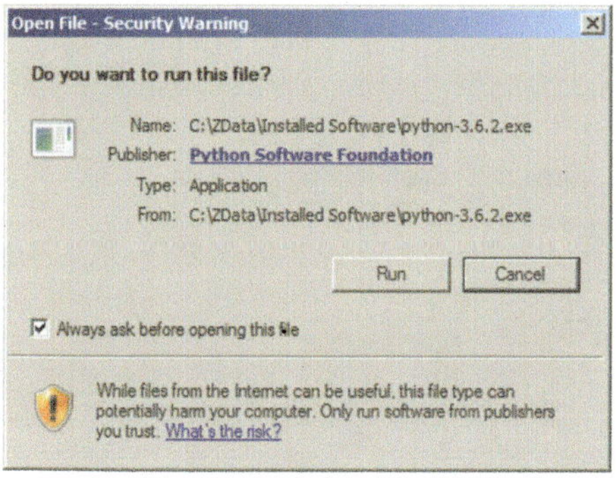

- Click Run.

  See below figure:

  Ensure that the Install launcher for all users (recommended) and the Add Python 3.6 to PATH checkboxes at the bottom are checked.

- Highlight the Install Now (or Upgrade Now) message, and then click it.

  A User Account Control pop-up window will appear, posing the question Do you want the allow the following program to make changes to this computer?

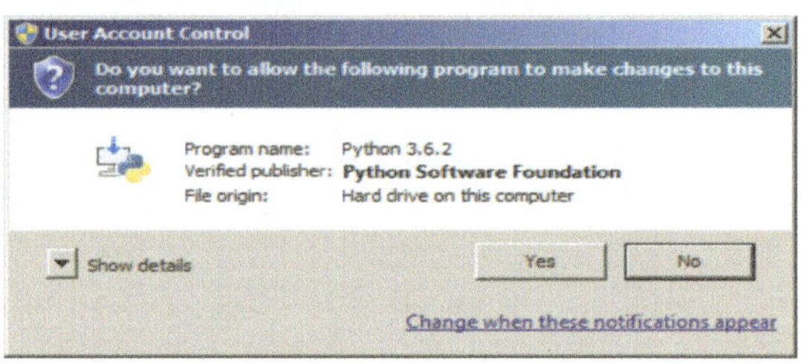

- Click the Yes button.

  A new Python 3.6.2 (32-bit) Setup pop-up window will appear with a Setup Progress message and a progress bar.

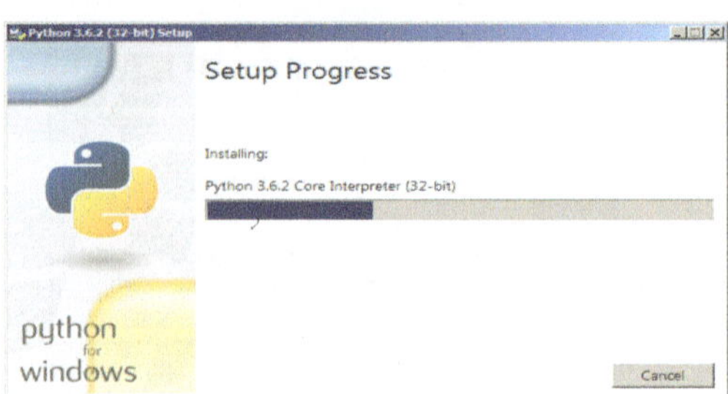

Setup was successful message will be displayed.

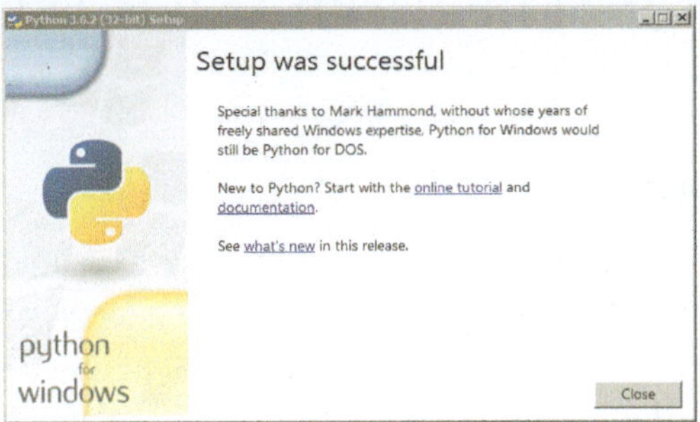

- Close and Python installed successfully.

4.2.3. Python on Linux

- Installing Python 3
- Installing Required Packages

To install Python, it should require prerequisites as shown below:

$ sudo apt-get install build-essential check install

To install supportive libraries, use the following command:

$ sudo apt-get install libreadline-gplv2-dev libncursesw5-dev libssl-dev libsqlite3-dev tk-dev libgdbm-dev libc6-dev libbz2-dev

To download python use the following commands:

$ cd /usr/src
$ sudo wget https://www.python.org/ftp/python/3.7.8/Python-3.7.8.tgz

Now extract the downloaded package as shown below:
$ sudo tar xzf Python-3.7.8.tgz

- **Compiling Python Source:** To compile Python source, use the following command:

$ cd Python-3.7.8

$ sudo./configure

To check the Python version, use the following command:

$ sudo python 3.7.8-V

The sample output should be like this:

Python 3.7.8

4.3. PYTHON INTERPRETER AND INTERACTIVE MODE

4.3.1. Interpreter

Python is an interpreted language because they are executed by an interpreter. Interpreter take high level program as input and executes what the program says. It processes the program a minimum at a time. It read lines and performs computations alternatively. Figure 4.1 explains the structure of an interpreter.

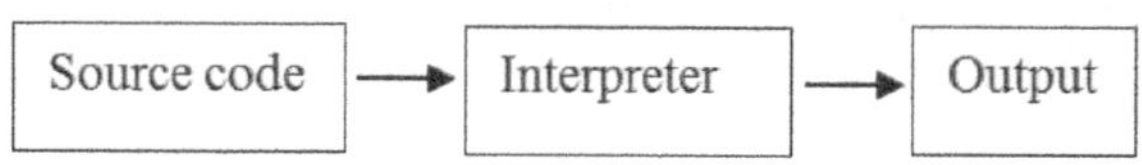

Figure 4.1. Function of interpreter.

4.3.2. Compiler

A compiler reads the program and interprets it to machine-readable form called object code or executable code before the program starts running. Once a program is compiled, the program can be executed repeatedly without further translations. Figure 4.2 shows the structure of a compiler.

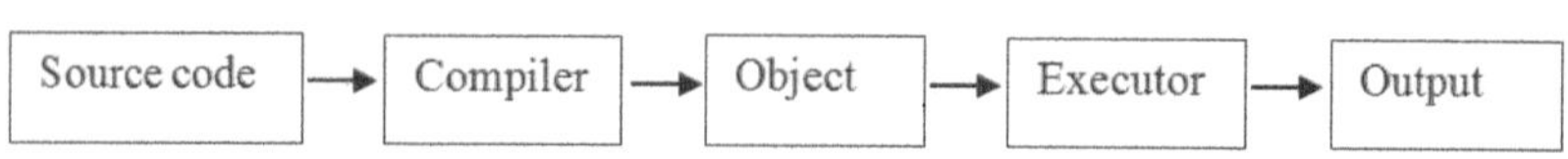

Figure 4.2. Function of compiler.

The python program that you have installed will by default act as something called an interpreter. An interpreter takes text commands and runs them as you enter them very handy for trying things out.

To execute the program code, the interpreter can be used. There are two different modes to use the interpreter. 1. Interactive mode, 2. Script mode. In interactive mode, the program statements can be typed in prompt, so the interpreter displays the result.

Just type python at your console, hit Enter, and you should enter Python's Interpreter. To find out which version of Python you're running, instead type python-V in your console to tell you.

Compiler	Interpreter
Compiler takes entire program as input	Interpreter takes single instruction as input
Intermediate object code is created	No intermediate object code is created
Conditional control statements are executes faster	Conditional control statements are executes slower
Requires more memory	Requires less memory
Program need not be compiled every time	Every time higher-level program is converted into lower-level program
Errors are displayed after entire program is checked	Errors are displayed for every instruction interpreted
Example: C Compiler	Example: Python

4.3.3. Interacting with Python

After opening Python, contextual information is shown:

```
Python 3.5.0 (default, Sep 20 2015, 11:28:25)
[GCC 5.2.0] on linux
Type “help,” “copyright,” “credits” or “license” for more information.
>>>
```

Now try this below code:

```
print(“Hello world”)
```

Press Enter. After viewing the results, Python goes to the interactive prompt, where you could enter another command:

```
>>> print(“Hello world”)
Hello world
>>> (1 + 4) * 2
10
```

4.3.4. Writing a Python Program

We will consider two ways:

- Enter the program directly into IDLE's interactive shell; and

Enter the program into IDLE's editor, save it, and run it.

1. **IDLE's Interactive Shell:** IDLE is a simple Python integrated development environment available for Windows, Linux, and Mac OS X. Figure 4.1 shows how to start IDLE from the Microsoft Windows Start menu. The IDLE interactive shell is shown in Figure 4.2. You may type the above one-line Python program directly into IDLE and press enter to execute the program. Figure 4.3 shows the result using the IDLE interactive shell.
2. **IDLE's Editor:** IDLE has a built-in editor. From the IDLE menu, select New Window, as shown in Figure 4.4. Type the text as shown in Listing 1.1 (simple.py) into the editor. Figure 4.5 shows the resulting editor window with the text of the simple Python program. You can save your program using the Save option in the File menu, as shown in Figure 4.6. Save the code to a file named simple.py. We can run the program from within the IDLE editor by pressing the F5 function key or from the editor's Run menu: Run→Run Module. The output appears in the IDLE interactive shell window (Figures 4.3–4.8).

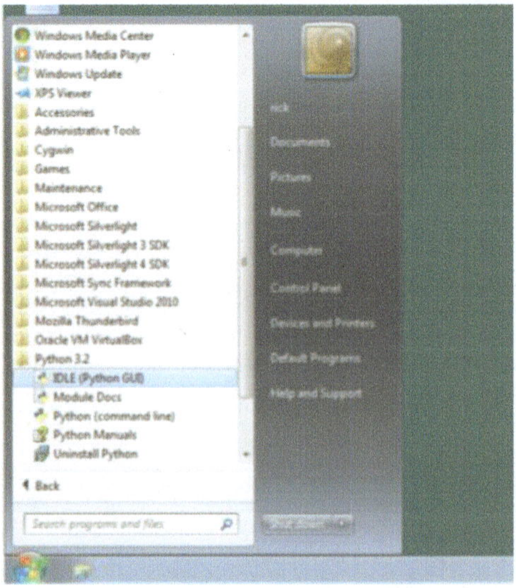

Figure 4.3. Start IDLE from the Windows Start menu.

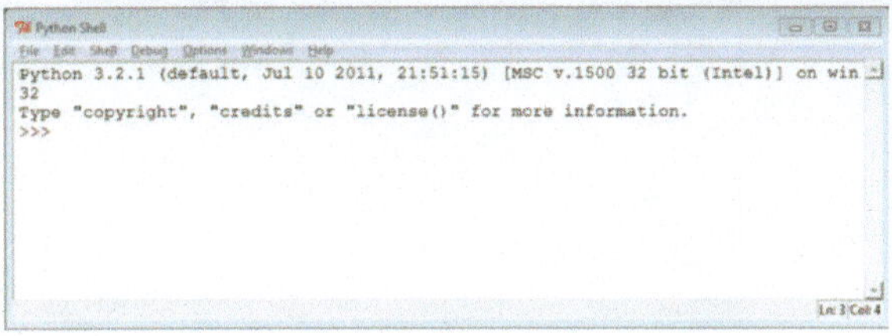

Figure 4.4. The IDLE interpreter Window.

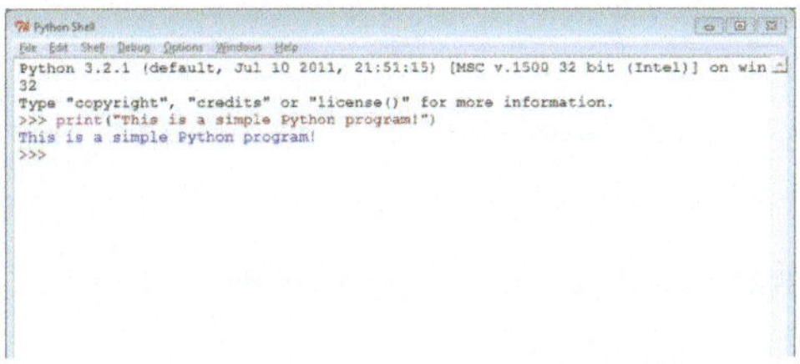

Figure 4.5. A simple Python program entered and run with the IDLE interactive shell.

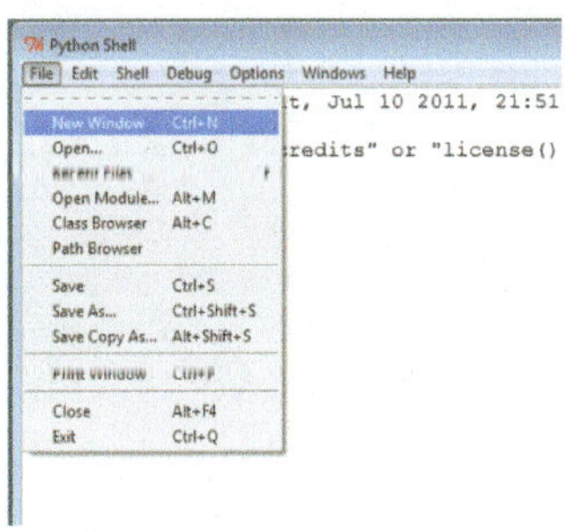

Figure 4.6. Launching the IDLE editor.

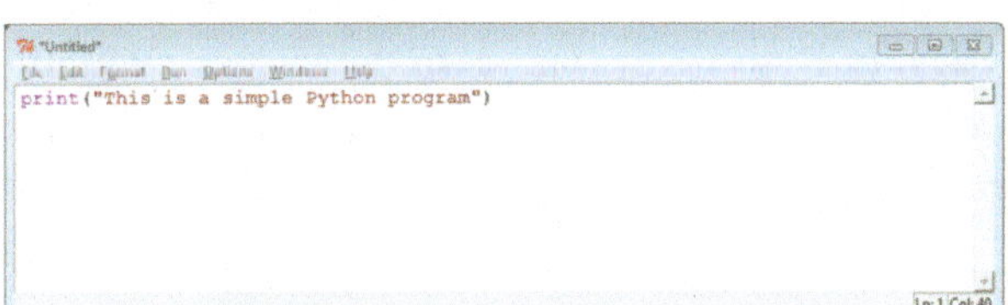

Figure 4.7. The simple Python program typed into the IDLE editor.

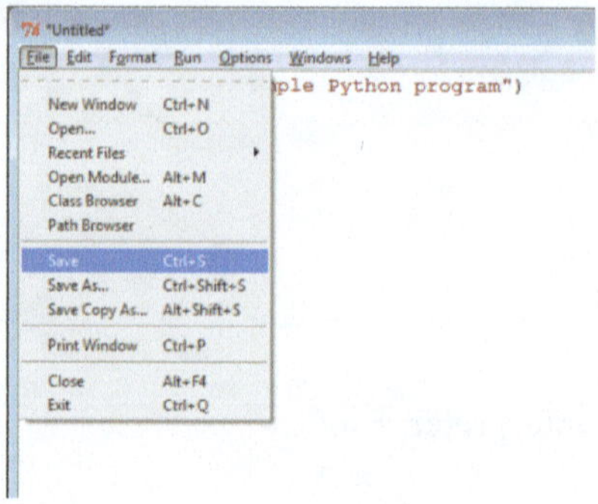

Figure 4.8. Saving a file created with the IDLE editor.

Example:

1. First simple program to print Hello, World!

    Print("Hello, World!")

It prints Hello, World. In Python, printing statement uses the keyword print as in the above-mentioned format. The parenthesis indicates that the print is a function. The single quotation represents the beginning and end of the text to be displayed.

4.4. VALUES AND TYPES: INT, FLOAT, BOOLEAN, STRING, AND LIST

Value:

A value is a basic thing that a program works with, like a letter or a number.

For example, the values are 2, 42.0, and 'Hello, World!.'
These values belong to different datatypes.

Data type:

Every value in Python has a data type. It is a set of values and the allowable operations on those values. The values belong to different data types. In Python, the standard data types available are:

- Numbers;
- String;
- List;
- Tuple;
- Dictionary.

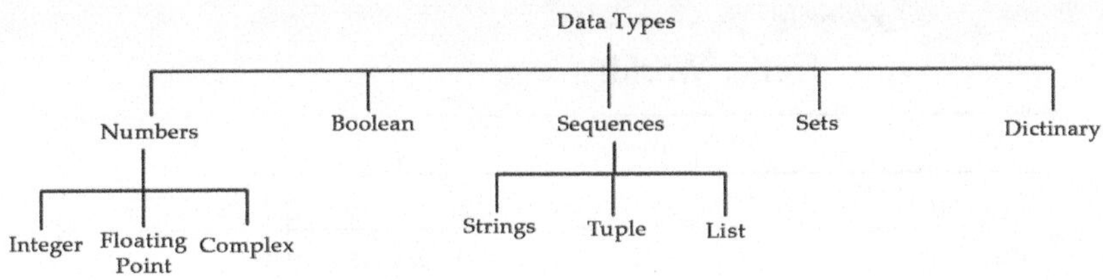

1. Number Data Type: These stores numerical values. It supports four numerical types. Number data type stores Numerical Values. This data type is immutable [i.e., values/items cannot be changed].

Python supports integers, floating-point numbers, and complex numbers. They are defined as:

i. **Integer:** They are positive or negative whole numbers without decimal points

   Example:

   signed numbers like 10, –20

   Long

   They are long integers. They can also be represented in octal and hexadecimal representation like, ox3245 and 234L

   Example:

   569243L

ii. **Float:** They are written with a decimal point dividing the integer and the fractional parts.

   Example:

   3.45

iii. **Complex:** Ordered pairs of real floating-point numbers represented by x ± jy, where x & y are real numbers and j is imaginary part.

   Complex numbers like, 7.32e-3j

2. **String Data Type:** Strings are the sequence of characters represented within quotation marks. It allows either pairs of single or double-quotes. The substring access is possible through the slicing operator ([] or [:]). The string index 0 represents the beginning of the string whereas, index-1 represents the ending of the string. The following examples illustrate the string and substring accesses.

Example:
str= "Hello, World!"

Code	Comment	Result
print(str)	# prints complete string	Hello, World!
print(str[0])	# prints first character of the string	H
print(str[-1])	#prints last character of the string	!
print(str[1:5])	# prints character starting from index 1 to 4 # prints str[start_at: end_at-1]	ello
print(str[2:])	#prints string starting at index 2 till end of the string	llo, World!
print(str * 2)	# asterisk (*)-is the repetition operator. Prints the string two times.	Hello, World!Hello, World!
print(str, 'Hai')	# prints concatenated string	Hello, World! Hai

3. **List Data Type:** Lists are the most significant compound data types that contain elements of various types. A List can hold items of different data types. The list is enclosed by square brackets [] where the items are separated by commas. Like string data type, the list values can be accessed using the slice operator.

1 or [:]). The index 0 represents the beginning of the list whereas, index-1 represents the ending of the list. The following example illustrates list accesses.

Example:

list1=['abcd,' 345, 3.2,'python,' 3.14]
list2=[234, 'xyz']

Code	Comment	Result
print(list1)	# prints complete list	['abcd,' 345, 3.2,'python,'
print(list1[0])	# prints first element of the list	abcd
print(list1[–1])	#prints last element of the list	3.14
print(list1[1:3])	# prints elements starting from index 1 to 2	[345, 3.2]

	# prints list1[start_at: end_at-1]	
print(list1[2:])	#prints list starting at index 2 till end of the list	[3.2, 'python,' 3.14]
print(list 2 * 2)	# asterisk (*)-is the repetition operator. Prints the list two times.	['abcd,' 345, 3.2, 'python,'
		3.14, 234, 'xyz']
print(list1 + list2)	# prints concatenated lists	['abcd,' 345, 3.2,'python,'
		3.14, 234, 'xyz']

4. **Tuple Data Type:** Tuple is another sequence data type similar to the list. A tuple consists of a number of values separated by commas and enclosed within parentheses. Unlike list, the tuple values cannot be updated. They are treated as read-only lists. The following example explains the tuple element access.

   Example:

   tuple1= ('abcd,' 345, 3.2,'python,' 3.14)
   tuple2= (234, 'xyz')

Code	Comment	Result
print(tuple1)	# prints complete tuple1	('abcd,' 345, 3.2, 'python,' 3.14)
print(tuple1[0])	# prints first element of the tuple1	abcd
print(tuple1[–1])	#prints last element of the tuple1	3.14
print(tuple1[1:3])	# prints elements starting from index 1 to 2	(345, 3.2)
	# prints tuple1[start_at: end_at-1]	
print(tuple1[2:])	#prints tuple1 starting at index 2 till the	(3.2, 'python,' 3.14)
	End	

print(tuple 2 * 2)	# asterisk (*)-is the repetition operator.	(234, 'xyz,' 234, 'xyz')
	Prints the tuple2 two times.	
print(tuple1 +	# prints concatenated tuples	('abcd,' 345, 3.2,'python,'
tuple2)		3.14, 234, 'xyz')

5. **Dictionary Data Type:** This is a kind of hash table. It contains key-value pairs. A dictionary key can be almost any Python type, usually numbers or strings. Values can be arbitrary Python object. Dictionaries are enclosed by curly braces {} and values can be assigned and accessed using square brackets []. The following example explains the dictionary element access.

```
dict1= {'name': 'ABCD,' 'code': 6734, 'dept': 'Engg'}
dict2= {}
dict2 ['rollno'] = "II-ITA24"
```

Code	Comment	Result
print(dict1)	# prints complete dictionary	{ 'dept': 'Engg,' 'code':6734, 'name': 'ABCD'}
print(dict1.keys())	# prints all keys of dictionary	{ 'dept,' 'code,' 'name'}
print(dict1.values)	#prints all values of dictionary	{'Engg,' 6734, 'ABCD'}
print(dict2['rollno'])	# print the value for the key rollno	II-ITA24

6. **Boolean Data Type:** Boolean is one more data type supported in Python. It takes the two values; True and False.

Example:

```
print(True)
        # True is a Boolean value
print(False)
        # False is a Boolean value
```

4.5. VARIABLES

A variable allows us to store a value by assigning it to a name, which can be used later. Named memory locations to store values. Programmers generally choose names for their variables that are meaningful. It can be of any length. No space is allowed. We don't need to declare a variable before using it. In Python, we simply assign a value to a variable, and it will exist.

4.5.1. Variable Declaration

In Python, interpreter automatically detects the type by the data to which it is assigned. For assigning values "=" is used.

Examples of variable declaration:

```
>>>X= 10# x is an integer
>>>Y=15.7                        # y is float
>>>Z="Welcome to Python"#        Z is a string
```

Assigning value to variable:

Value should be given on the right side of assignment operator(=) and variable on left side.

```
>>>counter =45
print(counter)
```

Assigning a single value to several variables simultaneously:

```
>>> a=b=c=100
```

Assigning multiple values to multiple variables:

```
>>> a,b,c=2,4,"ram"
```

4.6. KEYWORDS

Keywords are the reserved words in Python. We cannot use a keyword as variable name, function name or any other identifier. They are used to define the syntax and structure of the Python language. Keywords are case sensitive (Table 4.1).

Table 4.1. Python Keywords

and	Del		From		not	while
as	Elif		Global		or	with
assert	Else		If		pass	yield
break	Except		Import		print	
class	Exec		In		raise	
continue	Finally		Is		return	
def	For		Lambda		try	

4.7. STATEMENTS AND EXPRESSIONS

Statements:

Instructions that a Python interpreter can executes are called statements. A statement is a unit of code like creating a variable or displaying a value.

```
>>>   n = 17
>>>   print(n)
```

Here, The first line is an assignment statement that gives a value to n. The second line is a print statement that displays the value of n.

Expressions:

An expression is a combination of values, variables, and operators. A value all by itself is considered an expression, and also a variable. So the following are all legal expressions:

```
>>>   42
42
>>>   a=2
>>>   a+3+2
7
>>>   z=(“hi”+”friend”)
>>>   print(z)
hifriend
```

4.8. COMMENTS

Comments are the non-executable statements explain what the program does. For large programs, it often difficult to understand what it does. The comment can be added in the program code with the symbol #.

Example:

print('Hello, World!') # print the message Hello, World!; comment

v=5 # creates the variable v and assign the value 5; comment

4.9. INPUT AND OUTPUT

Input:

Input is data entered by user (end user) in the program. In python, input () function is available for input.

Syntax for input() is:

variable = input ("data")

Example:

>>> x=input("enter the name:") enter the name: george

>>>y=int(input("enter the number"))

enter the number 3

#python accepts string as default data type. conversion is required for type.

Output:

Output can be displayed to the user using a Print statement.

Syntax:

print(expression/constant/variable)

Example:

>>> print ("Hello") Hello

QUOTATION IN PYTHON:

>>>

Python accepts single ('), double (") and triple (''' or """) quotes to denote string literals.

>>>

Anything that is represented using quotations are considered as string.

- single quotes (' ')

E.g., 'This a string in single quotes'

- double quotes (“ ”)

  E.g., “‘‘This a string in double quotes’’”
- triple quotes(“““““ ”””””)

  E.g., This is a paragraph. It is made up of multiple lines and sentences.”””””

4.10. OPERATORS

An operator is a special symbol that asks the compiler to perform particular mathematical or logical computations like addition, multiplication, comparison, and so on. The values the operator is applied to are called operands. For example, in the expression 4 + 5, 4 and 5 are operands and + is an operator.

The following tokens are operators in Python:

<table>
<tr><td>+</td><td>–</td><td>*</td><td>**</td><td>/</td><td>//</td><td>%</td></tr>
<tr><td><<</td><td>>></td><td>&</td><td>|</td><td>^</td><td>~</td><td></td></tr>
<tr><td><</td><td>></td><td><=</td><td>>=</td><td>==</td><td>!=</td><td><></td></tr>
</table>

4.10.1. Types of Operator

Python language supports the following types of operators.

- Arithmetic operators;
- Comparison (relational) operators;
- Assignment operators;
- Logical operators;
- Bitwise operators;
- Membership operators;
- Identity operators;
- Unary arithmetic operators.

4.10.1.1. Arithmetic Operators

Operator	Description	Example
+ Addition	Adds values on either side of the operator.	a + b = 30
– Subtraction	Subtracts right hand operand from left hand operand.	a – b = –10
* Multiplication	Multiplies values on either side of the operator	a * b = 200
/ Division	Divides left hand operand by right hand operand	b / a = 2
% Modulus	Divides left hand operand by right hand operand and returns remainder	b % a = 0
** Exponent	Performs exponential (power) calculation on operators	a**b =10 to the power 20
// Floor Division	The division of operands where the result is the quotient in which the digits after the decimal point are removed	5//2=2

Examples	Output:
a=10	
b=5	
print("a+b=,"a+b)	a+b= 15
print("a-b=,"a-b)	a-b= 5
print("a*b=,"a*b)	a*b= 50
print("a/b=,"a/b)	a/b= 2.0
print("a%b=,"a%b)	a%b= 0
print("a//b=,"a//b)	a//b= 2
print("a**b=,"a**b)	a**b= 100000

4.10.1.2. Comparison (Relational) Operators

These are used to compare values. It either returns True or False according to the condition. Assume, a=10 and b=5.

Operator	Description	Example
==	If the values of two operands then the condition are equal, becomes true.	(a == b) is not true
!=	If the values of two operands are not equal, then condition becomes true.	(a!=b) is true
>	If the value of left operand is greater than the value of right operand, then condition becomes true.	(a > b) is not true.
<	If the value of left operand is less than the value of right operand, then condition becomes true.	(a < b) is true.
>=	If the value of left operand is greater than or equal to the value of right operand, then condition becomes true.	(a >= b) is not true.
<=	If the value of left operand is less than or equal to the value of right operand, then condition becomes true.	(a <= b) is true.

Example	Output
a=10	
b=5	
print("a>b=>,"a>b)	a>b=> True
print("a>b=>,"a<b)	a>b=> False
print("a==b=>,"a==b)	a==b=> False
print("a!=b=>,"a!=b)	a!=b=> True
print("a>=b=>,"a<=b)	a>=b=> False
print("a>=b=>,"a>=b)	a>=b=> True

4.10.1.3. Assignment Operators

These are used in Python to assign values to variables.

Operator	Description	Example
=	Assigns values from right side operand to left side operand.	c = a + b assigns value of a + b into c
+=	Performs addition using two operands and assigns the result to left side operand.	c += a is equivalent to c = c + a
-=	Subtracts right-side operand from the left side operand and assigns the result to left side operand.	c– = a is equivalent to c = c – a
*=	Performs multiplication using two operands and assigns the result to left side operand.	c*=a is equivalent to c = c * a
/=	Divides left side operand by the right side operand and assigns the result to left side operand.	c/=a is equivalent to c = c/a
%=	Finds modulus using two operands and assigns the result to left side operand.	c%=a is equivalent to c = c % a
**=	Performs exponential calculation and assigns the result to the left side operand.	c**=a is equivalent to c = c ** a
//=	Performs floor division and assigns the result to the left side operand	c//=a is equivalent to c = c // a

Example:

```
a = 21
b = 10
c = 0
c = a + b
print("Line 1-Value of c is," c)
c += a
print("Line 2-Value of c is," c)
c *= a
print("Line 3-Value of c is," c)
c /= a
print("Line 4-Value of c is," c)
c %= a
```

```
print("Line 5-Value of c is," c)
c **= a
print("Line 6-Value of c is," c)
c //= a
print("Line 7-Value of c is," c)
```

Output:

- **Line 1:** Value of c is 31
- **Line 2:** Value of c is 52
- **Line 3:** Value of c is 1092
- **Line 4:** Value of c is 52.0
- **Line 5:** Value of c is 2
- **Line 6:** Value of c is 2097152
- **Line 7:** Value of c is 99864

4.10.1.4. Logical Operators

These are and, or, not operators.

Operator	Description
And	Logical AND returns true, if, and only if both operands are true.
Or	Logical OR returns true, if any of the two operands is true.
Not	Logical NOT returns the logical negation of its operand.

Here, any nonzero number is interpreted as true and zero is interpreted as false. Both the and operator and the or operator expect two operands. not operator operates on a single operand.

The behavior of each logical operator is specified in a truth table for that operator.

The truth tables for and, or, and not (Tables 4.2–4.4).

Table 4.2. Truth Table of and Operator

Op1	Op2	Op1 and Op2
True	True	True
True	False	False
False	True	False
False	False	False

Table 4.3. Truth Table of or Operator

Op1	Op2	Op1 or Op2
True	True	True
True	False	True
False	True	True
False	False	False

Table 4.4. Truth Table of Not Operator

Op1	Not Op1
True	False
False	True

Example: Truth Function
a = True
b = False
print('a and b is,' a and b)
print('a or b is,' a or b)
print('not a is,' not a)
Output
a and b is False
a or b is True
not a is False

4.10.1.5. Bitwise Operators

This operates on one or more bit patterns at the level of individual bits (Table 4.5)

Example:

x = 10 (0000 1010 in binary)
y = 4 (0000 0100 in binary)

<table>
<tr><th>Operator</th><th>Description</th></tr>
<tr><td>&</td><td>Performs bitwise AND operation between two the operands.</td></tr>
<tr><td>|</td><td>Performs bitwise OR operation between two the operands.</td></tr>
<tr><td>^</td><td>Performs bitwise XOR (exclusive OR) operation between two the operands.</td></tr>
<tr><td>~</td><td>Performs bitwise 1's complement on a single operand.</td></tr>
<tr><td><<</td><td>Shifts the first operand left by the number of bits specified by the second</td></tr>
<tr><td></td><td>operand (bitwise left shift).</td></tr>
<tr><td>>></td><td>Shifts the first operand right by the number of bits specified by the second</td></tr>
<tr><td></td><td>operand (bitwise right shift).</td></tr>
</table>

<table>
<tr><th>a</th><th>B</th><th>a & b</th><th>a | b</th><th>a ^ b</th><th>~a</th></tr>
<tr><td>0</td><td>0</td><td>0</td><td>0</td><td>0</td><td>1</td></tr>
<tr><td>0</td><td>1</td><td>0</td><td>1</td><td>1</td><td>1</td></tr>
<tr><td>1</td><td>0</td><td>0</td><td>1</td><td>1</td><td>0</td></tr>
<tr><td>1</td><td>1</td><td>1</td><td>1</td><td>0</td><td>0</td></tr>
</table>

Table 4.5. *Truth Table of &, |, ^, ~ Operator*

a = 60	# 60 = 0011 1100
b = 26	# 13 = 0001 1010
c = a & b;	# 24 = 0001 1000
print("Result of Bitwise AND is," c)	

c = a | b;	# 62 = 0011 1110
print("Result of Bitwise OR is," c)	
c = a ^ b;	# 38 = 0010 0110
print("Result of Bitwise XOR is," c)	
c = ~a;	# –61 = 1100 0011

```
print("Result of Bitwise Ones Complement is," c)
c = a << 2;      # 240 = 1111 0000
print("Result of Bitwise Left Shift is," c)
c = a >> 2;      # 15 = 0000 1111
print("Result of Bitwise Right Shift is," c)
```

Sample output:

```
Result of Bitwise AND is 24
Result of Bitwise OR is 62
Result of Bitwise XOR is 38
Result of Bitwise Ones Complement is -61
Result of Bitwise Left Shift is 240
Result of Bitwise Right Shift is 15
```

4.10.1.6. Membership Operators

These test for membership in a sequence, such as strings, lists, or tuples and explained below.

Operator	Description
in	Evaluates to true if it finds a variable in the specified sequence and false otherwise.
not in	Evaluates to true if it does not finds a variable in the specified sequence and false otherwise

Sample code:

```
a = 6
b = 2
list = [1, 2, 3, 4, 5];
print(a in list)
print(a not in list)
print(b in list)
print(b not in list)
```

Sample output:

```
False
True
True
False
```

4.10.1.7. Identity Operators

They are used to check if two values (or variables) are located on the same part of the memory.

Identity operators compare the memory locations of two objects. They are explained below:

Operator	Description
is	Returns true if both operands point to the same object and false otherwise.
is not	Returns false if both operands point to the same object and true otherwise.

Sample code:

```
a = 20
b = 20
print(a is b)
print(id(a) == id(b))
print(a is not b)
b=30
print(a is b)
print(a is not b)
print(id(a) == id(b))
```

Sample output:

```
True
True
False
False
True
False
```

4.10.1.8. Unary Arithmetic Operators

Operator	Description
+	Returns its numeric argument without any change.
–	Returns its numeric argument with its sign changed.

Sample Code:

```
a = 10
b = +a
print(b)
c = –a
print(c)
```

Sample output:

```
10
–10
```

4.10.2. Operator Precedence

When more than one operator appears in an expression, the order of evaluation depends on the rules of precedence. For mathematical operators, Python follows mathematical convention. The acronym PEMDAS (parentheses, exponentiation, multiplication, division, addition, subtraction) is a useful way to remember the rules.

The following table summarizes the operator precedence in Python, from the highest precedence to the lowest precedence. Operators in the same box have the same precedence and group from left to right (except for comparisons, including tests, which all have the same precedence and chain from left to right and exponentiation, which groups from right to left).

Operator	**Description**	**Associativity**
(expressions...)	Binding or tuple display	left to right
[expressions...]	List-display	
{key: value...}	Dictionary display	
'expressions...'	String conversion	
x[index]	Subscription	left to right
x[index:index]	Slicing	

x(arguments...)	Call	
x.attribute	Attribute reference	
**	Exponentiation	right-to-left
+x	Unary plus	left to right
-x	Unary minus	
~x	Bitwise NOT	
*	Multiplication	left to right
/	Division	
//	Floor division	
%	Remainder	
+	Addition	left to right
-	Subtraction	
<<, >>	Bitwise Left Shift and Right Shift	left to right
&	Bitwise AND	left to right
^	Bitwise XOR	left to right
|	Bitwise OR	left to right
in, not in	Membership tests	Chain from left to right
is, is not	Identity tests	
<, <=, >, >=, <>, !=, ==	Comparisons	
Not	Boolean NOT	left to right
And	Boolean AND	left to right
Or	Boolean OR	left to right

Examples:

4 * (6–3) is 12, and
(1+2)**(6–3) is 27.
3**1+1 is 4, not 9.
2*1**4 is 2, not 16.
4*6–2 is 22, not 16.
4+2/2 is 5, not 3.
4/2*2 is 4, not 1.

- *Example 1:*
  a=9–12/3+3*2–1
  a=?

a=9–4+3*2–1
a=9–4+6–1
a=5+6–1
a=11–1 a=10

- *Example 2:*
a=2,b=12,c=1
d=a<b>c
d=2<12>1
d=1>1
d=False

- *Example 3:*
A=2*3+4%5–3/2+6
A=6+4%5–3/2+6
A=6+4–3/2+6
A=6+4–1+6
A=10–1+6
A=9+6
A=15

- *Example 4:*
a=2,b=12,c=1
d=a<b>c–1
d=2<12>1–1
d=2<12>0
d=1>0
d=True

- *Example 5:*
find m=?
m=–43|8&0|–2
m=–43|0|–2
m=–43|–2
m=–1

- *Example 6:*
a=2*3+4%5–3//2+6
a=6+4–1+6
a=10–1+6
a=15

CHAPTER 5

FUNCTIONS

CONTENTS

5.1. FUNCTION DEFINITION

Function is a sub program which consists of a set of instructions used to perform a specific task. A large program is divided into basic building blocks called function.

(i). Need for function:

- When the program is too complex and large, they are divided into parts. Each part is separately coded and combined into a single program. Each subprogram is called function.
- Debugging, Testing, and maintenance becomes easy when the program is divided into subprograms.
- Functions are used to avoid rewriting the same code again and again in a program.
- Function provides code re-usability
- The length of the program is reduced.

(ii). Types of function:

The functions can be classified into two categories:

- Built-in function and
- User defined function.

5.2. BUILT-IN FUNCTIONS

5.2.1. Built-In/Pre-Defined Function

- Built in functions means already created and stored functions in Python.
- These built in functions are always available for usage and accessed by a programmer. It cannot be modified.

print()	Print objects to the stream.
input()	Reads a line from input, converts it to a string (stripping a trailing newline), and returns that.
abs()	Return the absolute value of a number.
len()	Return the length (the number of items) of an object.

Built in Function	Description
>>>max(3,4) 4	# returns largest element
>>>min(3,4) 3	# returns smallest element
>>>len(“hello”) 5	#returns length of an object
>>>range(2,8,1) [2, 3, 4, 5, 6, 7]	#returns range of given values
>>>round(7.8) 8.0	#returns rounded integer of the given number
>>>chr(5) \x05’	#returns a character (a string) from an integer
>>>float(5) 5.0	#returns float number from string or integer
>>>int(5.0) 5	# returns integer from string or float
>>>pow(3,5) 243	#returns power of given number
>>>type(5.6) <type ‘float’>	#returns data type of object to which it belongs
>>>t=tuple([4,6.0,7]) (4, 6.0, 7)	# to create tuple of items from list
>>>print(“good morning”) Good morning	# displays the given object
>>>input(“enter name: ”) enter name: George	# reads and returns the given string

Example:

Program to find the ASCII value of the given character.

```
c = input(“Enter a character”)
print(“ASCII value of ,”c, “is,”ord(c))
```

```
Sample Input/Output:
p
ASCII value of p is 112
```

ord() function converts a character to an integer (ASCII value). It returns the Unicode code point of that character.

5.2.2. Type Conversion Functions

Python provides built-in functions that convert values from one type to another.

Function	Converting What to What	Example	
		>>> int('2014')	
int()	string, floating-point → integer	2014	
		>>> int(3.141592)	
		3	
		>>> float('1.99')	
float()	string, integer → floating point	1.99	
	Number	>>> float(5)	
		5.0	
		>>> str(3.141592)	
str()	integer, float, list, tuple, dictionary	'3.141592'	
	→ string	>>> str([1,2,3,4])	
		'[1, 2, 3, 4]'	
		>>> list('Mary')	# list of characters in
		'Mary'	
list()	string, tuple, dictionary → list	['M,' 'a,' 'r,' 'y']	
		>>> list((1,2,3,4))	# (1,2,3,4) is a tuple
		[1, 2, 3, 4]	

		>>> tuple('Mary')	
		('M,' 'a,' 'r,' 'y')	
tuple()	string, list → tuple	>>> tuple([1,2,3,4]) # [ ] for list, () for	
		tuple	
		(1, 2, 3, 4)	

>>> age = 21

>>> sign = 'You must be ' + age + 'Years old'

'+' can also be used for concatenation, but Many Python functions are sensitive to the type of data. For example, you cannot concatenate a string with an integer. If you try, it will result in the following error.

Traceback (most recent call last):

File "<pyshell#71>," line 1, in <module> sign = 'You must be ' + age + 'years old' TypeError: cannot concatenate 'str' and 'int' objects

For the example above, use the str() conversion function to convert integer to string data type.

age = 21

sign = "You must be " + str(age) + "Years old"

>>>sign

Sample output:

You must be 21 Years old

Examples using Built-in functions for type conversion:

Program Code			
	Output		
Converting float to int			
>>>print(3.14, int(3.14))	3.14	3	
>>>print(3.9999, int(3.9999))	3.9999		3
>>>print(3.0, int(3.0))	3.0	3	
>>>print(–3.999, int(–3.999))	–3.999		–3
Converting string to int			

>>>print("2345," int("2345"))	2345	2345
>>>print(int("23bottles"))	Error:	
	ValueError: invalid literal for int() with	
	base 10: '23bottles'	
Converting int to string		
>>>print(str(17))	17	
Converting float to string		
>>>print(str(123.45))	123.45	
>>>print(type(str(123.45)))	<class 'str'>	
Converting list to tuple		
>>>fruits = ['apple,' 'orange,' 'grapes,'		
'pineapple']	('apple,' 'orange,' 'grapes,' 'pineapple')	
>>>print(tuple(fruits))		
	('P,' 'y,' 't,' 'h,' 'o,' 'n')	
>>>print(tuple('Python'))		
Converting tuple to list		
>>>print(list('Python'))	['P,' 'y,' 't,' 'h,' 'o,' 'n']	

5.3. MATH FUNCTIONS

Math and cmath are mathematical modules available in Python to support familiar mathematical functions. A module is a file that contains a collection of related functions. Before using built-in math functions, import math module.

```
>>>import math
```

It will create a module object named math which contains functions and variables defined in the module. Some of the familiar math functions are listed in the table.

Function	Description	Example	Output
abs(n)	Return the absolute value of a number(n)	abs(–99)	99
round(n,d)	Round a number(n) to a number of decimal points(d)	round(3.1415,2)	3.14
floor(n)	Round down to nearest integer	math.floor(4.7)	4.0
ceil(n)	Round up to nearest integer	math.ceil(4.7)	5.0
pow(n,d)	Return n raised to the power d	math.pow(10,3)	1000.0
sqrt(n)	Returns the square root of number(n)	math.sqrt(256)	16.0
fsum(iterable)	Return an accurate floating point sum of values in the iterable	sum([.1,.1,.1,.1,.1,.1,.1,.1,.1,.1]) fsum([.1,.1,.1,.1,.1,.1,.1,.1,.1,.1])	0.99999999999999 1.0
factorial(n)	Return n factorial	math.factorial(5)	120
gcd(n,m)	Return greatest common division of (n,m)	math. gcd(10,125)	5
trunc(x)	Return the real value x truncated to an integral	math. trunc(1.999)	1

sin(x), cos(x), tan(x)	Return the arc sine, cosine, tangent of x, in radians.	math.sin(math.pi/4)	0.7071067811865476
		math.cos(math.pi)	–1.0
		math.tan(math.pi/6)	0.5773502691896257

5.4. USER DEFINED FUNCTION

The functions defined by the users according to their requirements are called user-defined functions. The users can modify the function according to their requirements.

Advantages of user-defined functions:

- Large project are divided into smaller different functions.
- Repeated code could be included and executed.

➢ **Function Definition: (Sub Program)**

- A header-starts with def (keyword) and ends with a colon.
- A body contains one or more Python statements each indented the same amount-4 spaces is the Python standard-from the header. Inside the function add the program statements to be executed End with or without return statement

Syntax:

```
def fun_name(Parameter1,Parameter2…Parameter n):
        statement1
        statement2…
        statement n
        return[expression]
```

The code block inside every function commences with a colon (:) and is indented. The first statement of a function can be an optional statement-the documentation string of the function or docstring. The statement return [expression] exits a function, and it returns the result of the function. The rules for function names are the same as for variable names: letters, numbers, and some punctuation marks are legal, but the first character can't be a number. Keywords should not be used as function name. To avoid confusion, use different names for functions and variables.

Example:

```
def my_add(a,b):
        c=a+b
```

```
return c
```

➤ **Function Calling: (Main Function):**

A function can be executed by calling it from another function or directly from the Python prompt by its name.

Syntax:

```
function_name(parameters)
```

Function to display Welcome message.

```
def display():
print("Welcome!!!")
>>>display()
```

Sample output:

```
Welcome!!!
```

The first line of the function definition is called the header; the rest is called the body. The header has to end with a colon, and the body has to be indented. By convention, the indentation is always four spaces. In this example, the function name is display(). The empty parentheses after the name indicate that this function doesn't take any arguments.

Function to find the biggest of three numbers.

```
def great(no1,no2,no3):
    if (no1>no2) and (no1>no3):
        return no1
    elif (no2>no3):
        return no2
    else:
        return no3
n1=int(input("Enter first number"))
n2=int(input("Enter second number"))
n3= int(input("Enter third number"))
result=great(n1,n2,n3)
print(result, "is bigger")
```

Sample input/output:

```
Enter first number 10
Enter second number 5
Enter third number 25
25 is bigger
```

In this example, the function name is great().It takes three parameters and returns the greatest of three. input() reads input value from the user. The function is invoked by calling great(n1,n2,n3). print() displays the output. The strings in the print statements are enclosed in double-quotes.

Flow of execution:

- The order in which statements are executed is called the flow of execution;
- Execution always begins at the first statement of the program;
- Statements are executed one at a time, in order, from top to bottom;
- Function definitions do not alter the flow of execution of the program, but remember that statements inside the function are not executed until the function is called;
- Function calls are like a bypass in the flow of execution. Instead of going to the next statement, the flow jumps to the first line of the called function, executes all the statements there, and then comes back to pick up where it left off.

5.5. FUNCTION PROTOTYPES

1. **Function Without Arguments and Without Return Type:** In this type, no argument is passed through the function call, and no output is return to the main function.

   The subfunction will read the input values perform the operation and print the result in the same block.

Example:

```
def add():
        a=int(input("Enter a"))
        b=int(input("Enter b"))
        c=a+b
        print(c)
add()
```

Output:

```
Enter a 5
Enter b 10
15
```

2. **Function with Arguments and Without Return Type:** Arguments are passed through the function call, but output is not return to the main function.

Example:

```
def add(a,b):
        c=a+b
        print(c)
a=int(input("Enter a"))
b=int(input("Enter b"))
add(a,b)
```

Output:

```
Enter a 5
Enter b 10
15
```

3. **Function Without Arguments and with Return Type:** In this type, no argument is passed through the function call, but output is return to the main function.

Example:

```
def add():
        a=int(input("Enter a"))
        b=int(input("Enter b"))
        c=a+b
        return c
c=add()
print(c)
```

Output:

```
Enter a 5
Enter b 10
15
```

4. **Function with Arguments and with Return Type:** In this type, arguments are passed through the function call and output is return to the main function.

Example:

```
def add(a,b):
        c=a+b
return c
        a=int(input("Enter a"))
b=int(input("Enter b"))
c=add(a,b)
print(c)
```

Output:

```
Enter a 5
Enter b 10
15
```

5.5.1. Parameters and Arguments

1. ***Parameters:*** These are the value(s) provided in the parenthesis when we write function header. These are the values required by function to work. If there is more than one value required, all of them will be listed in the parameter list separated by comma.

   Example:

   ```
   def my_add(a,b):
   ```

2. ***Arguments:*** These are the value(s) provided in the function call/invoke statement. List of arguments should be supplied in the same way as parameters are listed. Bounding of parameters to arguments is done 1:1, and so there should be the same number and type of arguments as mentioned in the parameter list.

   Example:

   ```
   my_add(x,y)
   ```

Arguments types:

i. **Required Arguments:** The number of arguments in the function call should match exactly with the function definition.

```
def my_details(name, age):
        print("Name:," name)
        print("Age," age)
        Return
```

```
my_details("george,"56)
```

Output:

```
Name: george
Age: 56
```

ii. **Keyword Arguments:** Python interpreter is able to use the keywords provided to match the values with parameters even though if they are arranged in out of order.

```
def my_details(name, age):
        print("Name:," name)
        print("Age," age)
        return
my_details(age=56,name="george")
```

Output:

```
Name: george
Age: 56
```

iii. **Default Arguments:** Assumes a default value if a value is not provided in the function call for that argument.

```
def my_details(name, age=40):
print("Name:," name)
print("Age," age)
return
my_details(name="george")
```

Output:

```
Name: george
Age: 40
```

iv. **Variable Length Arguments:** If we want to specify more arguments than specified while defining the function, variable length arguments are used. It is denoted by * symbol before parameter.

```
def my_details(*name):
        print(*name)
my_details("rajan,""rahul,""micheal," "ärjun")
```

Output:

```
rajan rahul micheal ärjun
```

5.6. RETURN STATEMENT

The return statement is used to exit a function and go back to the place from where it was called. If the return statement has no arguments, then it will not return any values. But exits from function.

Syntax:

```
return[expression]
```

Example:

```
def my_add(a,b):
        c=a+b
        return c
x=5
y=4
print(my_add(x,y))
```

Output:

```
9
```

5.7. MODULES

A module is a file containing Python definitions, functions, statements, and instructions. Standard library of Python is extended as modules. To use these modules in a program, the programmer needs to import the module.

Once we import a module, we can reference or use to any of its functions or variables in our code.

- There is a large number of standard modules also available in Python.
- Standard modules can be imported the same way as we import our user-defined modules.
- Every module contains many function.
- To access one of the function, you have to specify the name of the module and the name of the function separated by dot. This format is called dot notation.

Syntax:

```
import module_name
module_name.function_name(variable)
```

Importing built-in module:

```
import math
x=math.sqrt(25)
print(x)
```

Importing user defined module:

```
import cal
x=cal.add(5,4)
print(x)
```

Built-in python modules are:

1. Math-mathematical functions: Some of the functions in math module is:

math.ceil(x)	Return the ceiling of x, the smallest integer greater than or equal to x
math.floor(x)	Return the floor of x, the largest integer less than or equal to x.
math.factorial(x)	Return x factorial.
math.gcd(x,y)	Return the greatest common divisor of the integers a and b
math.sqrt(x)	Return the square root
math.log(x)	Return the natural logarithm of x
math.log10(x)	Returns the base-10 logarithms
math.log2(x)	Return the base-2 logarithm of x.
math.sin(x)	Returns sin of x radians
math.cos(x)	Returns cosine of x radians
math.tan(x)	Returns tangent of x radians
math.pi	The mathematical constant π = 3.141592
math.e	Returns The mathematical constant e = 2.718281

2 .random-Generate Pseudo-Random Numbers

```
random.randrange(stop)
random.randrange(start, stop[, step])
random.uniform(a, b)
```

Return a random floating point number

5.7.1. Illustrative Programs

1. Program for SWAPPING (exchanging) of values:

```
a = int(input("Enter a value "))
b = int(input("Enter b value "))
c = a
a = b
b = c
print("a=,"a,"b=,"b,)
```

Output:

```
Enter a value 5
Enter b value 8
a=8
b=5
```

2. Program to find distance between two points:

```
import math
x1=int(input("Enter x1"))
y1=int(input("Enter y1"))
x2=int(input("Enter x2"))
y2=int(input("Enter y2"))
distance =math.sqrt(((x2-x1)**2)+((y2 – y1)**2))
print(distance)
```

Output:

```
Enter x1 7
Enter y1 6
Enter x2 5
Enter y2 7
2.36
```

3. Program to circulate n numbers:

```
a=list(input("Enter the list"))
print(a)
for i in range(1,len(a),1):
        print(a[i:]+a[:i])
```

Output:

```
Enter the list '1234'
```

```
['1,' '2,' '3,' '4']
['2,' '3,' '4,' '1']
['3,' '4,' '1,' '2']
['4,' '1,' '2,' '3']
```

4. Python program to test for leap year:

Note: A leap year is divisible by 4, but not by 100, unless it is divisible by 400.

```
def leapyr(yr):                                          # function definition
    if yr%4==0 and yr%100!=0 or yr%400==0:               #condition-leap year
        print("Leap Year")
    else:
        print("Not a Leap Year")
    return
year=int(input("Enter a year"))
leapyr(year)                                             # function call
```

Sample input/output:

```
Enter a year1900
Not a Leap Year
```

5. Python function to check whether a number is in a given range:

```
    def test_range(n):
    if n in range(1,10):
        print("No. is between 1 and 10")
    else:
        print ("No. is not between 1 and 10")
no=int(input("Enter a no."))
test_range(no)
```

Sample input/output:

```
Enter a no. 10
    No. is not between 1 and 10
Enter a no. 4
```

```
No. is between 1 and 10
```

6. Python program to print the even numbers from a given list:

```
def even(l):
    for n in l:
        if n % 2 == 0:
            print('\n,' n)
    return
even([1, 2, 3, 4, 5, 6, 7, 8, 9])
```

Sample output:

```
2
4
6
8
```

7. Function that circulates/rotates the list values for given number of times (order):

```
def rotate(l,order):
    for i in range(0,order):
        j=len(l)–1
        while j>0:
            tmp=l[j]
            l[j]=l[j–1]
            l[j–1]=tmp
            j=j–1
        print (i, 'rotation,'l)
    return
l=[1,2,3,4,5]
rotate(l,3)
```

Sample output:

```
0 rotation: [5, 1, 2, 3, 4]
1 rotation: [4, 5, 1, 2, 3]
2 rotation: [3, 4, 5, 1, 2]
```

8. Function that takes a number and check the number is prime or not:
Note: A prime number (or a prime) is a natural number greater than 1 and that has no positive divisors other than 1 and itself.

```
def test_prime(n):                          # function that returns Boolean value
        if(n==1):
                return False
        elif(n==2):
                return True
        else:
                for x in range(2,n):
                        if(n%x==0):
                                return False
                return True
no=int(input("Enter a number"))
if(test_prime(no)):
        print(no, "is Prime")
else:
        print(no, "is not Prime")
```

Sample output:

```
Enter a number9
9 is not Prime
Enter a number11
11 is Prime
```

9. Function to check whether a number is perfect or not:

Example: The first perfect number is 6, because 1, 2, and 3 are its proper positive divisors, and 1 + 2 + 3 = 6. Equivalently, the number 6 is equal to half the sum of all its positive divisors: (1 + 2 + 3 + 6) / 2 = 6.

```
def perfect_number(n):
        sum = 0
        for x in range(1, n):
                if n % x == 0:
                        sum += x
        return sum
no=int(input("Enter a number"))
sum=perfect_number(no)
if(sum==no):
                        print("Perfect number")
```

```
else:
        print("Not a Perfect number")
```

Sample input/output:

```
Enter a number5
Not a Perfect number
Enter a number6
Perfect number
```

10. Function that checks whether a number is palindrome or not:

```
def Palindrome_Number():
    no =int(input("Enter a Number:"))
    q=no
    rev = 0
    while(q!=0):    # loop for reverse
        rev = (q % 10) + (rev * 10)
        q = q // 10
    if(rev == no):
        print("%d is a palindrome number" %no)
    else:
        print("%d is not a palindrome number" %no)
Palindrome_Number()
```

Sample input/output:

```
Enter a Number121
121 is a palindrome number
```

11. Find the distance between two points (x_c,y_c) and (x_p,y_p):

```
import math
def distance(x1, y1, x2, y2):
    dx = x2 – x1
    dy = y2 – y1
    dsquared = dx**2 + dy**2
    result = math.sqrt(dsquared)
    return result
xc=int(input("Enter xc"))
yc=int(input("Enter yc"))
```

```
xp=int(input("Enter xp"))
yp=int(input("Enter yp"))
print (distance(xc,yc,xp,yp))
```

Sample input/output:

```
Enter xc 2
Enter yc 2
Enter xp 4
Enter yp 4
2.82
```

Control Flow Statements

CHAPTER 6

CONTROL STRUCTURES

CONTENTS

6.1. BOOLEAN VALUES

6.1.1. Boolean

The Boolean values are True and False. The relational operators such as ==, !=, >, <, >=, <= and the logical operators such as and, or, not are the Boolean operators. The statement that prints either true or false is a Boolean expression. Given below is the example of the operator != that inspects two operands and produces True if they are not equal else displays False.

```
6.      5 != 6
True
7.      5 != 5
False
```

So, True, and False are special values which belong to the Boolean type; they are not strings (case sensitive):

To know the type of data, the following example can be used.

```
6.      type(True)
<type 'bool'>
7.      type(False)
<type 'bool'>
```

The relational operators are as follows.

Boolean function works with relational operator, string comparison and logical operators.

6.1.1.1. Relational Operators

Relational operators compares values and evaluate single value either True or False. The relational operators are as follows:

```
a == b  #if a is equal to b
a != b  # if a is not equal to b
a > b   # if a is greater than b
a < b   #if a is less than b
a >= b  # if a is greater than or equal to b
a <= b  #if a is less than or equal to b
```

Simple program to explain relation operators for Boolean conditions:

```
a = 10
b = 12
print('a > b is,'a>b)
print('a < b is,'a<b)
```

```
print('a == b is,'a==b)
print('a != b is,'a!=b)
print('a >= b is,'a>=b)
print('a <= b is,'a<=b)
```

The result is:

```
a > b is False
a < b is True
a == b is False
a != b is True
a >= b is False
a <= b is True
```

There are three logical operators: and, or, and not.
For example;

- **and operator:**

(a > 0) and (a < 10)

- True only if a is greater than 0 and less than 10.
- Otherwise False

- **or operator:**

(n%2 == 0) or (n%3 == 0)

- True if either of the conditions is true, that is, if the number is divisible by 2.
- Otherwise False

- **not operator:**
  - not operator negates a Boolean expression
  - not (a > b)
  - True if a > b is False, that is, if a is less than or equal to b.
  - False if a>b is True, that is if a is greater than b

6.2. CONDITIONAL STATEMENTS

6.2.1. Condition Statement (if)

Conditional statements provide the ability to check conditions and control the program execution accordingly. The simplest form of conditional statement is the if statement. The syntax of the if statement is given below.

Syntax:

```
if test expression:
        statement(s)
        (or)
if test expression: statement
```

Flowchart:

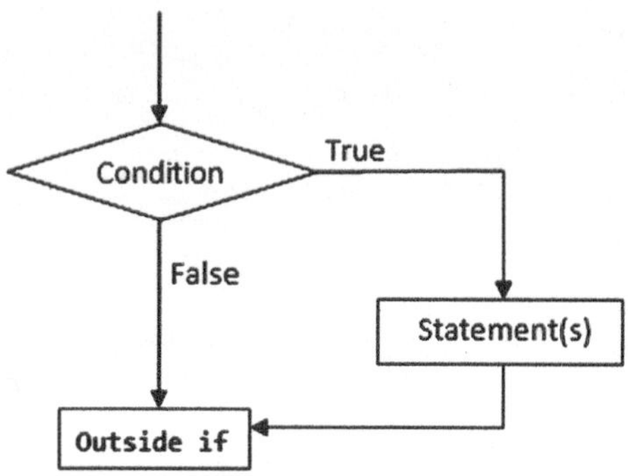

The program evaluates the test expression and will execute statement(s) only if the text expression is True. If the text expression is False, the statement(s) is not executed.

Example:

```
if x > 0:
        print('x is positive')
```

Program code:

```
num = 5
if (num%2) != 0:
        print(num, "is odd number.")
print("This is always printed.")
num = 4
if (num%2) == 0:
        print(num, "is even number.")
print("This is also always printed.")
```

Result:

```
5 is an odd number.
        This is always printed.
4 is even number.
        This is also always printed.
```

6.2.2. Alternative Execution (If...Else Statement)

The if…else statement is called alternative execution, in which there are two possibilities and the condition determines which one gets executed. The syntax of if…else statement is given below.

Syntax:

```
if test expression:
        Body of if
else:
        Body of else
```

The if...else statement evaluates the test expression and will execute the body of if only when test condition is True. And if the condition is False, body of else is executed. Indentation is used to separate the blocks.

Flowchart:

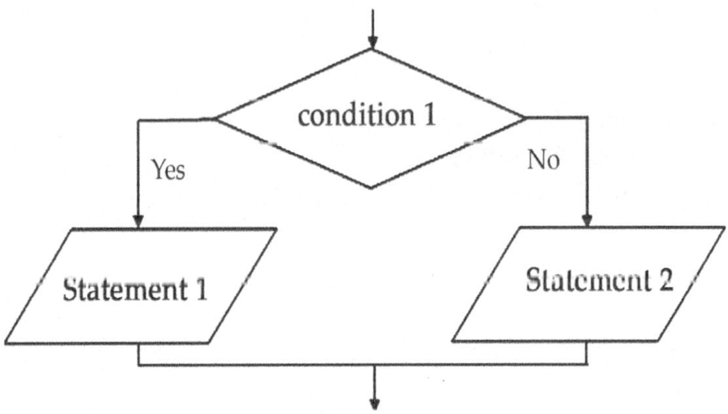

- *Example 1:* Odd or even number

```
n=int(input("Enter a number"))
if(n%2==0):
        print("Even number")
else:
        print("Odd number")
```

Output:

```
Enter a number4
Even number
```

- *Example 2:* Positive or Negative Number

```
n=int(input("Enter a number"))
```

```
if(n>=0):
        print("Positive number")
else:
        print("Negative number")
```

Output:

```
Enter a number8
Positive number
```

- *Example 3:* greatest of two numbers

```
a=int(input("Enter a value:"))
b=int(input("Enter b value:"))
if(a>b):
        print("Greatest:,"a)
else:
        print("Greatest:,"b)
```

Output:

```
enter a value:4
enter b value:7
greatest: 7
```

6.2.3. Chained Conditionals (If-Elif-Else)

Chained conditionals (if-elif-else) allows more than two possibilities and need more than two branches. The syntax of if-elif-else is shown below.

Syntax:

```
if test expression:
        Body of if
elif test expression:
        Body of elif
else:
        Body of else
```

Flowchart:

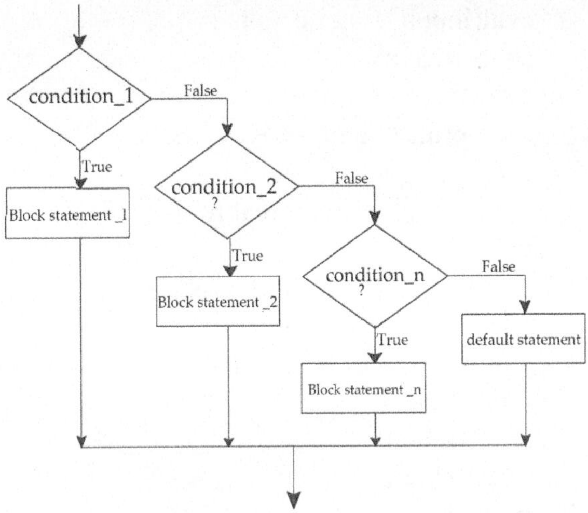

The example explains the if-elif-else:

```
if x < y:
        print('x is less than y')
elif x > y:
        print('x is greater than y')
else:
        print('x and y are equal')
```

Here, elif stand for "else if." There can be numerous elif statements and at least one branch will be executed. The program of the elif statements will be terminated by an else statement at the end.

- *Example 1:* Positive, Negative or Zero

```
num = 5.4
if num > 0:
        print("Positive number")
elif num == 0:
        print("Zero")
else:
        print("Negative number")
```

Output:

```
Positive number
```

- *Example 2:* Roots of quadratic equation

```
a=eval(input("Enter a value:"))
b=eval(input("Enter b value:"))
```

```
c=eval(input("Enter c value:"))
d=(b*b–4*a*c)
if(d==0):
        print("Same and real roots")
elif(d>0):
        print("Diffrent real roots")
else:
        print("Imaginagry roots")
```

Output:

```
Enter a value:1
Enter b value:0
Enter c value:0
Same and real roots
```

- *Example 3:* Greatest among Three Numbers

```
a = int(input("Enter first number: "))
b = int(input("Enter second number: "))
c = int(input("Enter third number: "))
if (a > b) and (a > c):
        largest = a
elif (b > a) and (b > c):
        largest = b
else:
        largest = c
print("The largest number between,"a,,""b,"and,"c,"is,"lar
gest)
```

Output:

```
Enter first number: 4
Enter second number: 56
Enter third number: 6
('The largest number between,' 4.0,,'' 56.0, 'and,' 6.0, 'is,'
56.0)
```

6.2.4. Nested Conditionals

Any number of conditional statements can be nested inside one another. To indicate the level of nesting, indentation is used. The structure of nested conditionals is shown below.

```
if test expression:
        Body of if
else:
        if test expression:
        Body of if
else:
        if test expression:
        Body of if
:
:
else:
        Body of else
```

Example:

```
x=3
y=4
if x == y:
        print('x and y are equal')
else:
        if x < y:
        print('x is less than y')
else:
        print('x is greater than y')
```

In this program, the variables x and y are assigned with values 3 and 4 respectively. In the first If statement it checks whether x is equal to y. If it is true, then prints x and y are equal. If it is false, it executes the else part. Here, the else part contains the if statement (Nested if) checks whether x is lesser than y. If it is true, then it prints x is less than y. If it is false, then it prints x is greater than y that is the statement in else part.

Flowchart:

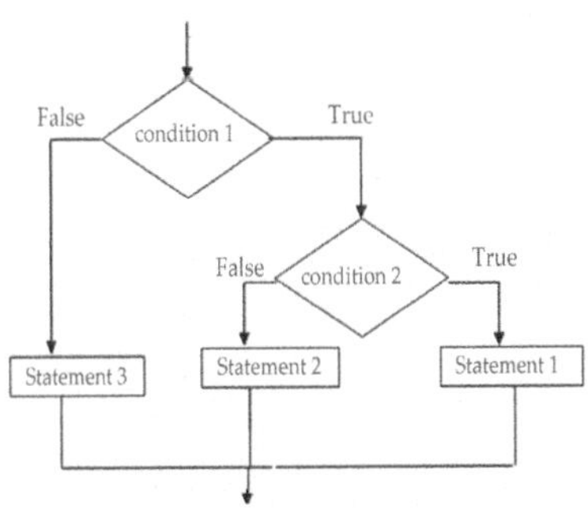

Example 1: Greatest of three numbers

```
a=int(input("Enter the value of a"))
b= int (input("Enter the value of b"))
c= int (input("Enter the value of c"))
if(a>b):
        if(a>c):
                print("The greatest no is,"a)
        else:
                print("the greatest no is,"c)
else:
        if(b>c):
                print("the greatest no is,"b)
        else:
                print("the greatest no is,"c)
```

Output:

```
Enter the value of a 9
Enter the value of a 1
Enter the value of a 8
The greatest no is 9
```

6.3. ITERATION/CONTROL STATEMENTS

6.3.1. State of a Variable

It is possible to have more than one assignment for the same variable. The value which is assigned at the last is given to the variable. The new assignment makes an existing variable assigned with a new value by replacing the old value.

For example, consider the following multiple assignments.

```
x=5
y=3
x=4
print x
print y
```

The result is:

```
4
3
```

The variable x is first assigned with 5 then it is assigned with 4. The last assignment statement x=4 replaces the old value of x (x=5).

Consider the following program code for multiple assignments.

```
x = 5
y = a    # x and y are now equal
x = 3    # x and y are no longer equal
print x
print y
```

The result is:

```
3
5
```

6.3.2. Looping Statements

Normally, program statements are executed sequentially, one after another. In certain criteria, it is necessary to execute a certain number of blocks repeatedly. These are repetitive program codes, the computers have to perform to complete tasks. The computers are used to automate the repetitive tasks. Programming languages present various control structures like looping statements which permit for more difficult implementation paths. The function of the looping statements is to execute a set of loops repeatedly (iterations). The following types of loops are used in Python programming language to handle looping requirements.

Loop Type	Description
While loop	This loop starts with condition checking. It repeats the execution of a statement or group of statements while the given condition is TRUE. Every time it tests the condition before executing the loop body.
For loop	It executes a statement or a group of statements multiple times and abbreviates the code that manages the loop variable.
Nested loops	One or more loops used in another loop.

1. **While Loop:** The function of the while loop is to execute repeatedly until the given condition becomes false. First and foremost, the test condition will be confirmed in the while loop. When the test condition is true, the program enters into the body of the loop. Subsequent to the completion of iteration, the test condition will be checked again. This process persists until the test_expression becomes False.

In Python, the body of the while loop is decided through indentation. The statements inside the while start with indentation and the first unindented line marks the end.

Syntax:

```
while test_expression:
        Body of while
```

Flowchart:

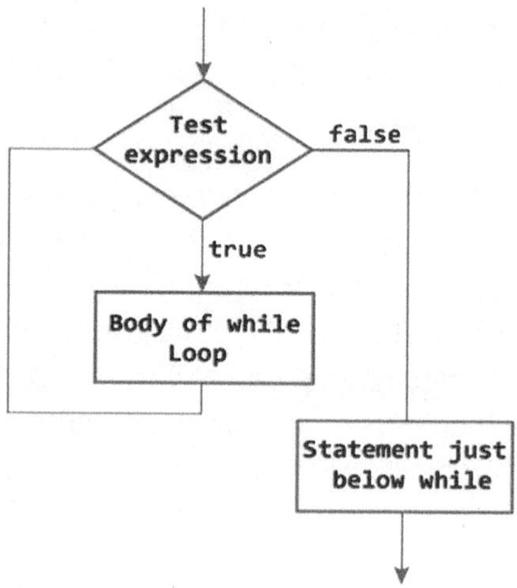

Example:

```
count = 0
while count < 5:
        print(count)
        count += 1
```

The result is:

```
0
1
2
3
4
```

- *Example 1:* Program to add natural numbers up to n

```
n = int(input("Enter n: "))
sum = 0
i = 1
```

```
while i <= n:
        sum = sum + i
        i = i+1
print ("The sum is," sum)
```

Output:

```
Enter n: 10
The sum is 55
```

- *Example 2:* Factorial of a numbers:

```
n=int(input("Enter n"))
i=1
fact=1
while(i<=n):
        fact=fact*i
        i=i+1
print(fact)
```

Output:

```
Enter n 5
120
```

- *Example 3:* Sum of digits of a number

```
n=int(input("enter a number"))
sum=0
while(n>0):
        a=n%10
        sum=sum+a
        n=n//10
print(sum)
```

Output:

```
enter a number
123
6
```

- *Example 4:* Reverse the given number

```
n=int(input("enter a number"))
sum=0
```

```
while(n>0):
        a=n%10
        sum=sum*10+a
        n=n//10
print(sum)
```

Output:

```
enter a number
123
321
```

- *Example 5:* Armstrong number or not

```
n=int(input("enter a number"))
org=n
sum=0
while(n>0):
        a=n%10
        sum=sum+a*a*a
        n=n//10
if(sum==org):
        print("The given number is Armstrong number")
else:
        print("The given number is not Armstrong number")
```

Output:

```
enter a number153
        The given number is Armstrong number
```

- *Example 6:* Palindrome or not

```
n=int(input("Enter a number"))
org=n
sum=0
while(n>0):
        a=n%10
        sum=sum*10+a
        n=n//10
if(sum==org):
        print("The given no is palindrome")
else:
```

```
print("The given no is not palindrome")
```

Output:

```
Enter a number121
        The given no is palindrome
```

2. **For Loop:** Python uses For loop to iterate over a sequence of elements (list, tuple, string) or other iterable objects. Iterating over a sequence of items is named as traversal. The syntax of for loop is shown below.

Syntax:

```
for val in sequence:
        Body of for
```

The val is the loop variable which takes the value of the item inside the sequence on each iteration. The loop continues until the last element is reached in the sequence. The body of for loop is marked using indentation.

Flowchart:

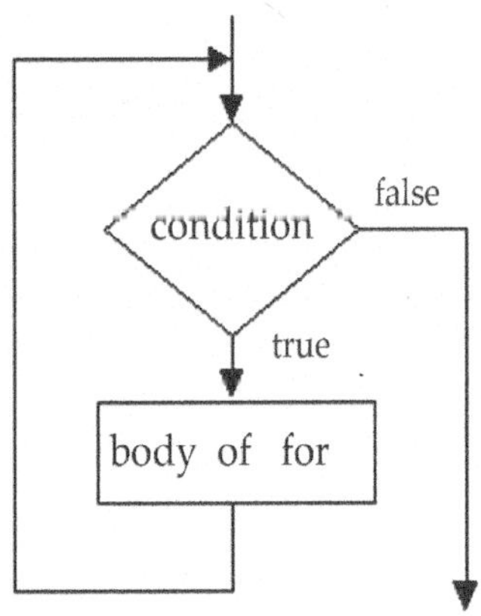

- *Example 1:*

```
numbers = [3, 2, 5, 7, 9, 1, 4, 6, 8]
total= 0
for item in numbers:
        total = total+item
print ("The total is," total)
```

Result:

```
The total is 45
```

Example 1: Fibonacci series

```
a=0
b=1
n=eval(input("Enter the number of terms: "))
print("Fibonacci Series: ")
print(a)
print(b)
for i in range(1,n,1):
    c=a+b
    print(c)
    a=b
    b=c
```

Output:

```
Enter the number of terms: 6
Fibonacci Series:
0
1
1
2
3
5
8
```

- *Example 2:* Check the no is prime or not

```
n=eval(input("Enter a number"))
for i in range(2,n):
if(n%i==0):
        print("The num is not a prime")
        break
else:
        print("The num is a prime number.")
```

Output:

```
Enter a no:7
The num is a prime number.
```

- *Example 3:* check a number is perfect number or not

```
n=int(input("enter a number:"))
sum=0
for i in range(1,n,1):
        if(n%i==0):
                sum=sum+i
if(sum==n):
        print("the number is perfect number")
else:
        print("the number is not perfect number")
```

Output:

```
enter a number:6
the number is perfect number
```

- *Example 4:* Program to print first n prime numbers

```
number=int(input("Enter no of prime numbers to be dis-
played:"))
count=1
n=2
while(count<=number):
        for i in range(2,n):
                if(n%i==0):
                        break
        else:
                print(n)
                count=count+1
        n=n+1
```

Output:

```
enter no of prime numbers to be displayed:5
2
3
5
7
11
```

- *Example 5:* Program to print prime numbers in range

```
lower=int(input("Enter a lower range"))
upper=int(input("Enter a upper range"))
```

```
for n in range(lower,upper + 1):
if n > 1:
        for i in range(2,n):
                if (n % i) == 0:
                        break
        else:
                print(n)
```

Output:

```
Enter a lower range50
Enter a upper range100
53
59
61
67
71
73
79
83
89
97
```

6.4. LOOP CONTROL STATEMENTS

Loop control statements change execution from its normal sequence. When execution leaves a scope, all automatic objects that were created in that scope are destroyed. Python supports the following control statements.

Control Statement	Description
Break statement	It terminates or breaks the loop statement and transfers the flow of execution to the statement immediately following the loop.
Continue statement	Causes the loop to skip the rest of its body and immediately retest its condition before reiterating.
Pass statement	The pass statement in Python is used when a statement is required syntactically but you do not want any command or code to execute.

6.4.1. Break Statement

The break statement is used to change or alter the flow of execution. The looping statements iterates till the test expression is true. In some cases, the current iteration of the loop need to be terminated. The break statement is used in this case. If the break statements are used in the nested loops, then it will terminate the particular loop where it has been used.

Syntax of break:

Break

Flowchart:

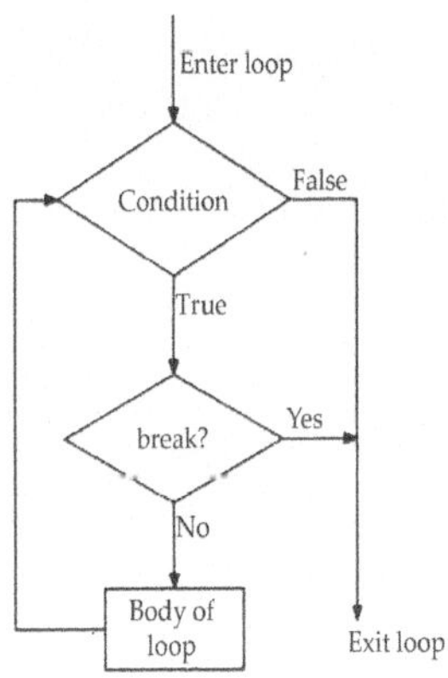

The working of break statement in for loop is explained as follows:

1. For var in sequence:

if condition: # it is true; executes break

break

codes inside the loop

#codes outside the loop

2. While test expression:

#codes inside the loop

if condition: # it is true; executes break

break

codes inside the loop

#codes outside the loop

Example:

```
for i in "welcome":
        if(i=="c"):
                break
        print(i)
```

Output:

```
        w
        e
        l
```

6.4.2. Continue Statement

The continue statement will skip the rest of the code inside a loop for the current iteration without terminating the loop.

Syntax:

```
        Continue
```

Flowchart:

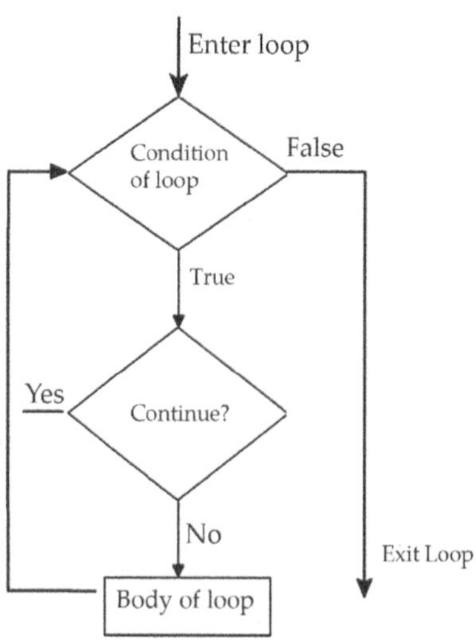

The working of continue statement in while loop is shown below.
While test expression:

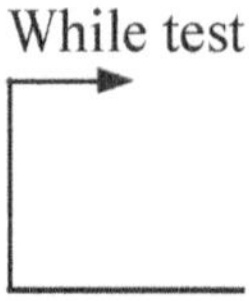

#codes inside the loop

```
if condition:    # it is true; executes
continue
#codes inside the loop
#codes outside the loop
```

Example:

```
for i in "welcome":
        if(i=="c"):
                continue
        print(i)
```

Output:

```
w
e
l
o
m
e
```

6.4.3. PASS

Pass statement executes nothing. It results in No operation. In Python programming, pass is a null statement. There is a slight difference between comment and pass statement. The comment statement ignores a comment entirely by the interpreter, but this will not happen in pass statement.

Syntax of pass:

```
pass
```

The pass statement can be used in places where the program code cannot be left as blank. But that can be written in future. The pass is used as placeholders. Pass is used in to construct program codes that do nothing.

- *Example 1:*

```
for i in "welcome":
        if(i=="c"):
                pass
        print(i)
```

Output:

```
w
e
l
o
m
```

e

- *Example 2:*
  sequence = {'p,' 'a,' 's,' 's'}
  for val in sequence:
  pass

The result is:

no output will be displayed

Difference between break and continue

Break	Continue
It terminates the existing loop and executes the residual statement outside the loop.	It terminates the current iteration and transfer the control to the next iteration in the loop
Syntax: Break	Syntax: continue

for i in "welcome":
if(i=="c"):
break
print(i)
	for i in "welcome":
	if(i=="c"):
	continue
	print(i)

w e l	w e l c o m e

Else statement in loops:

Else in for loop:

In this case, the else will be executed when the loop reaches a limit. The statements inside for loop and statements inside else will also execute.

Example:

```
for i in range(1,6):
        print(i)
else:
        print("the number greater than 6")
```

Output:

1

```
2
3
4
5
The number greater than 6
```

Else in while loop:

In this case, when the while loop becomes false, the else loop will be executed. The statements inside for loop as well as else will also execute.

```
Program
        i=1
        while(i<=5):
        print(i)
        i=i+1
else:
        print("the number is greater than 5")
```

Output:

```
1
2
3
4
5
the number greater than 5
```

6.5. FRUITFUL FUNCTIONS

Fruitful functions are functions that return value. While using fruitful function, the return value must be handled properly by assigning it to a variable or use it as part of the expression.

```
import math
x=math.sin(90)+1
print(x)                              # Output is 1.8939966636
```

In a script, calling a fruitful function without assigning the return value will result in loss.

```
import math
math.sin(90)                          # no output-as the return value is not assigned
```

6.5.1. Void Functions

Void function is a function that always returns None. It represents the absence of value.

```
def show():
        print('Welcome!!!')
result=show()
print(result)
```

Output:

```
Welcome!!!
None
```

Return values:

Return keywords are used to return the values from the function.

Example:

return a-return 1 variable
return a,b– return 2 variables
return a,b,c– return 3 variables
return a+b– return expression
return 8– return value

6.6. LOCAL AND GLOBAL SCOPE

6.6.1. Global Scope

The scope of a variable refers to the places that you can see or access a variable. A variable with global scope can be used anywhere in the program. It can be created by defining a variable outside the function.

6.6.2. Local Scope

A variable with local scope can be used only within the function.

Example:

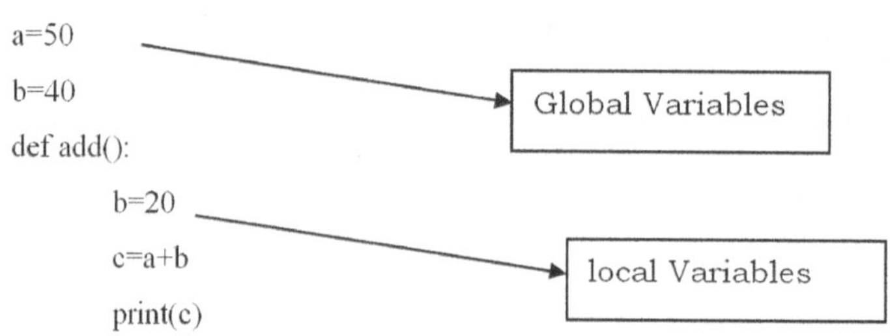

```
def sub():
        b=30
        c=a-b
        print(c)
print(a)
add()
sub()
print(b)
```

Output:

```
50
70
20
40
```

Note: If a variable has been defined both inside and outside a function, then local value will be taken inside the function, and global value will be taken outside the function.

6.7. FUNCTION COMPOSITION

Function composition is the ability to call one function from within another function. It is a way of combining functions such that the result of each function is passed as the argument of the next function. In other words, the output of one function is given as the input of another function is known as function composition.

Example: Compute area of circle with the given inputs center point (xc,yc) and perimeter point (xp,yp).

Note: First find the radius of circle by computing distance between (xc,yc) and (xp,yp). Later find the area of circle using radius.

```
import math
def distance(x1, y1, x2, y2):
        dx = x2 – x1
        dy = y2 – y1
        dsquared = dx**2 + dy**2
        result = math.sqrt(dsquared)
        return result
def area(radius):
        return (math.pi*radius*radius)
xc=int(input("Enter xc"))
yc=int(input("Enter yc"))
```

```
xp=int(input("Enter xp"))
yp=int(input("Enter yp"))
print (area(distance(xc,yc,xp,yp)))
```

Output:

```
Enter xc10
Enter yc10
Enter xp15
Enter yp10
78.5398163397
```

6.8. RECURSION

Recursion is a way of programming in which a function calls itself again and again until a condition is true. A recursive function calls itself and has a termination condition.

- **Advantages of Recursion:**
  - Recursive functions provide a good look to the program.
  - With the help of the recursive function, we can break a complex program into a simple one.
  - Rather than using iteration, recursive function provides a simple way for the sequence generation.
- **Disadvantages of Recursion:**
  - At certain situation, it is very hard to follow the recursion logic.
  - Recursive calls consume a lot of memory spaces that makes it inefficient.
  - Recursive functions are very tough to debug.

6.8.1. Program for Factorial Computation Using Recursion

The process involved in factorial computation using recursion. For example 5! computation can be represented as:

5!=5*4!; 4!=4*3!; 3!=3*2!; 2!=2*1!; 1!=1*0!; and 0!=1

Factorial (5)

↑

Return 5 * Factorial (4) =120

↑

Return 4 * Factorial (3) = 24

↑

Return 3 * Factorial (2) = 6

↑

Return 2 * Factorial (1) =2

↑

Return 1 * Factorial (0) =1

```
        1
def factorial(n):
if n ==0:          # base case
return 1
else:
return n * factorial(n – 1)
n=int(input("Enter a number:"))
print (factorial(n))
```

Output:

```
        Enter a number:5
        120
```

Example 1: sum of n numbers using recursion

```
        def sum(n):
                if(n==1):
                return 1
        else:
                return n+sum(n–1)
        n=int(input("Enter no.sum:"))
        sum=sum(n)
        print("Fact is,"sum)
```

Output:

```
Enter no. to find sum:10
Fact is 55
```

CHAPTER 7

STRINGS

CONTENTS

7.1. STRING DEFINITION

A string is a sequence of null or any number of characters. An empty string which holds no characters and has length 0.The indices of a string's characters are numbered from 0 from the left end and numbered from –1 from the right end. String is defined as a sequence of characters correspond to in quotation marks (either single quotes (') or double quotes (").

For a string 'WELCOME,' it's index values are shown in the following table:

Character	W	E	L	C	O	M	E
Index from left end	0	1	2	3	4	5	6
Index from right end	–7	–6	–5	–4	–3	–2	–1

7.1.1. Subscript Operator

[ ] is a subscript operator that is used to access the characters one at a time.

Syntax:

< stringname>[<index>]

Index within [ ] indicates the position of the particular character in the string, and it must be an integer expression.

Sample Code to print H

```
s= "HELLO"
print(s[0])
```

7.2. OPERATIONS ON STRING

1. Indexing:

String A	H	E	L	L	O
Positive Index	0	1	2	3	4
Negative Index	-5	-4	-3	-2	-1

```
>>>a="HELLO"
```

```
>>>print(a[0])
>>>H
>>>print(a[–1])
>>>O
```

Positive indexing is used to access the string from the beginning. Negative subscript helps in access the string from the end.

2. Slicing: A part of a string is called a slice. The operator [m:n] returns the part of the string from mth index to nth index, including the character at mth index but excluding the character at nth index. By omitting the first index, slicing will happen from the beginning of the string. By omitting the second index, the slice moves to the last part of the string. If the first index is greater than or equals to the second, the slice is an empty string. If both indices are omitted, the slice is a given string itself.

```
print[0:4]-HELL
print[:3]-HEL
print[0:]-HELLO
```

3. Concatenation: The + operator joins the text on both sides of the operator.

```
a="save"
b="earth"
>>>print(a+b)
saveearth
```

4. Repetitions: The * operator repeats the string on the left side number of times the value on right hand side.

```
a="panimalar "
>>>print(3*a)
panimalarpanimalarpanimalar
```

5. Membership: A membership operator compares two strings with the Boolean operator "in." If the particular string is present in the substring, then it returns true, else returns false. Usually membership operator is used to check whether a particular character is present in the string or not.

```
>>> s="good morning"
>>>"m" in s
True
>>> "a" not in s
True
```

7.2.1. Len Function

Len is a built-in function. It returns the number of characters in a string.

Sample code:

```
s= "HELLO PYTHON!"
print("Length is," len(s))
```

Output:

```
Length is 13
Immutability
```

The string is an immutable data structure. This means that its characters can be accessed, but the string cannot be modified.

Example:

```
s='hello, Python!'
s[0]='H'
TypeError: 'str' object does not support item assignment
```

7.3. STRING METHODS

Python consists of the following built-in methods for manipulating strings:

S L . No.	Method	Description
1.	s.capitalize()	It capitalizes only the first letter of a given string s
2.	s.center(width, fillchar)	It returns a space-padded string where a given string s is centered within the specified width.
3.	s.count(substr)	Return the number of occurrences of substr within a given string s
4.	s.endswith(substr)	Boolean function that returns True, if a string s ends with substr. Else, returns False.
5.	s.find(substr)	Returns the smallest index in s, if substr is found within s. Else, returns –1.
6.	s.index(substr)	Similar to find(), but raises an exception if substr is not found.

7.	s.isalnum()	Boolean function that returns True, if s is nonempty and all characters are alphanumeric. Else, returns False.
8.	s.isalpha()	Boolean function that returns true, if s is nonempty and all characters are alphabet. Else, returns False.
9.	s.isdigit()	Boolean function that returns true if string is nonempty and all characters are digits. Else, returns False.
10.	s.islower()	Boolean function that returns true, if s is nonempty and all characters are in small letters. Else, returns False.
11.	s.isupper()	Boolean function that returns true, if s is not an empty and all characters are in capital letters. Else, returns False.
12.	sp.join(seq)	Returns a concatenation of the set of strings in the given sequence (seq). The separator between the elements is sp.
13.	len(s)	It returns the number of characters of a string
14.	s.lower()	Returns a lowercase letters of the given string.
15	max(s)	Returns the biggest alphabetical character from the string s.
16.	min(s)	Returns the smallest alphabetical character from the string s.
17.	s.replace(old, new)	Returns a copy of s after replacing all occurrences of a substring old by a substring new.
18.	s.split()	Returns a list of the words from the string, using any whitespace as a delimiter.
19.	s.startswith(substr)	Boolean function. It returns True, when a string s starts with substring. Else, returns False.
20.	s.swapcase()	Returns all characters where they are case-inverted.
21.	title()	Returns all the characters which are all in starting capital letter
22.	s.upper()	Returns all characters in Uppercase.

23.	s.istitle()	Boolean function that returns True, When a string is in title case. Else, returns False.

1. capitalize(): This function returns a string with the first letter capitalized. It doesn't modify the old string.

Syntax:

string.capitalize()

The capitalize() function doesn't take any parameter.

Example:

```
Str1 = "python is awesome."
str2 = str1.capitalize()
print("Old String:," str1)
print("Capitalized String:,"str2)
```

Output:

```
Old String: python is awesome.
Capitalized String: Python is awesome.
```

2. center(): This method which returns a string that is padded with the particular character.

The syntax of center() method is:

string.center(width[, fillchar])

The center() method takes two arguments:

- width-It is the length of the string with padding characters
- fillchar (optional)-padding character

Example:

```
str = "Python is good"
new = str.center(24)
print("Centered String:," new)
```

Output:

```
Centered String: Python is good
```

3. count(): This method of the string takings the number of incidence of a substring in the given string.

Normally it can be said as the count () method will search for the substring in a given string and will return the number of times the substring has occurred in the string. There is also the optional parameters, start, and end which is to point out the starting and ending positions in the string in that order.

Syntax:

stringname.count(substring, start=..., end=...)

count() method only necessitates a solitary parameter for execution. Though, it also has two optional parameters:

i. **Substring:** It is a string to be counted
ii. **Start (Optional):** It is the starting position from where the search to begin
iii. **End (Optional):** It is the end position up to which position we want to search

Example:

```
string1 = "Python is awesome, isn't it?"
substring = "is"
count = string1.count(substring)
print("The count is:," count)
```

Output:

```
The count is: 2
```

4. endswith(): This method returns true if the given string is ended with the specified substring otherwise it will returns False

Syntax:

```
stringname.endswith(suffix[, start[, end]])
```

The endswith() have three parameters:

i. **Suffix:** It is the String or tuple of suffixes to be tartan.
ii. **Start (Optional):** It is a Beginning location where suffix is to be checked inside the string.
iii. **End (Optional):** It is an Ending location where suffix is to be checked inside the string.

Program:

```
text = "Welcome to Python to learn."
Result1 = text.endswith('to learn')
print(Result1)
Result1 = text.endswith('to learn.')
print(Result1)
Result 1= text.endswith("Welcome to Python to learn.")
print(Result1)
```

Output:

```
False
True
True
```

5. find(): This method is used to find the given substring in the given sentence. If it is not found it will return –1

Syntax:

```
stringname.find(sub[, start[, end]])
```

The find() method takes maximum of three parameters:

i. **Sub:** It's the substring to be looked for in the str string.
ii. **Start and End:** substring is searched within str[start:end]. It is the optional one

The find() method returns an integer value.

- If substring is survived inside the string, it will produce the lowest index where the substring is originate.
- If substring does not exist inside the string, it returns –1.

Program:

```
qe = "Hello Python"
result1 = qe.find('Hello')
print("Substring 'Hello':," result1)
result1 = qe.find('welcome')
print("Substring 'welcome':," result1)
if (qe.find('be,') != –1):
        print("It Contains substring 'be,'")
else:
        print("Doesn't contain substring")
```

Output:

```
Substring 'Hello': 0
Substring 'welcome ': –1
Doesn't Contains substring
```

6. index(): The index() function is used to find the string position which is mentioned inside of the index(). If the substring is not found, then the exception will be raised.

Syntax:

string variable. Index(sub[, start[, end]])

The index() method acquires three parameters:

- **Sub:** It is a substring to be searched in the string str.
- **Start and End(Optional):** It is a substring is searched surrounded by str[start:end]

Return value of the index()

- If the substring is found in the given string then the index of the first matched substring will be returned
- The value error exception will be raised when the substring is not found.

Program:

```
String1 = 'Welcome to Python programming.'
Result1= String1.index('Python')
print("Substring 'Python':," Result1)
Result2 = String1.index('Java')
print("Substring 'Java':," Result2)
```

Output:

```
Substring 'Python': 12
Result2 = String1.index('java')
***
***
ValueError: substring not found
```

7. isalnum(): This method isalnum()returns True when all characters in the string are alphanumeric (It should be either alphabets or numbers). Otherwise it returns False.

Syntax:

string.isalnum()

The isalnum() doesn't take any parameters.

The isalnum() returns:

True if all characters in the string are alphanumeric

False if at least one character is not alphanumeric

Program:

```
name = "Hello4"
```

```
print(name.isalnum())
name = "Hello4 world6"
print(name.isalnum())
name = "Hello4world6"
print(name.isalnum())
name = "655"
print(name.isalnum())
```

Output:

```
True
False
True
True
```

8. isalpha(): This method produces True if all characters in the string are alphabets. Otherwise, it returns False.

Syntax:

stringname.isalpha()
There is no parameter inside of isalpha()
The Output of isalpha() is:
True-if all characters in the string are alphabets (both lowercase and uppercase acceptable).
False-if at least one character is not alphabet.

Program:

```
n = "Hello"
print(n.isalpha())
n = "22145"
print(n.isalpha())
n= "Hello Wor33ld"
print(name.isalpha())
```

Output:

```
True
False
False
```

9. islower(): This method proceeds True if all alphabets in a string are lowercase alphabets. If the string holds at least one uppercase alphabet, it precedes False.

Syntax:

```
stringname.islower()
```

The islower() method do not take any parameters.

The islower() method returns:

- It Returns True if all alphabets that exist in the string are lowercase alphabets.
- It Returns False if the string contains at least one uppercase alphabet.

Program:

```
st = 'hello world'
print(st.islower())
st= 'hello5 worl3d'
print(st.islower())
st = 'hello World'
print(st.islower())
```

Output:

```
True
True
False
```

7.4. STRING MODULE

The string module of Python is a file that offers additional functions, classes, and variables to manipulate standard strings. But, some methods that are available in the standard data structure are not found in the string module (For example, isalpha()).

When we import a module, the following syntax is used:

```
import module1[, module2[,... modulen]
```

When an 'import' statement is encountered by the interpreter, the corresponding module(s) is imported if it is available in the search path.

dir() Function

It returns a sorted list of strings that includes the names of all modules, functions, and variables that are defined in a module.

Sample code using dir():

```
import string
```

```
content = dir(string)
print(content)
```

Sample output:

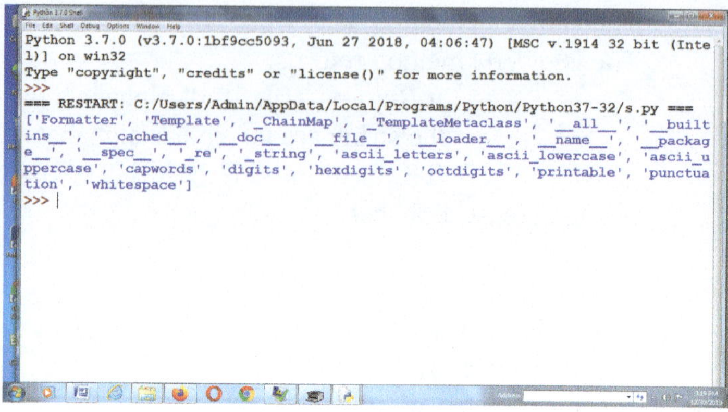

Escape sequences

Escape Sequence	Description	Example
\n	new line	>>> print("hai \nhello") hai hello
\\	prints Backslash (\)	>>> print("hai\\hello") hai\hello
\'	prints Single quote (')	>>> print("\'") '
\"	prints Double quote (")	>>>print("\"") "
\t	prints tab sapace	>>>print("hai\thello") hai hello
\a	ASCII Bell (BEL)	>>>print("\a")

7.5. LIST AS ARRAY

Python lists can store standards of dissimilar data types. But, arrays in python can only store values of similar data type. Array is not a fundamental data type in Python. So, the standard 'array' module has to be imported as:

```
from array import *
```

Then an array has to be declared as:

```
arrayID = array(typecode, [Initializers])
```

Here, 'arrayID' is the array name, 'typecode' is the array type and 'Initializers' are the values with which an array is initialized.

Example:

```
my_array = array('i,'[1,2,3,4])
```

Type Code	Description
'b'	It is a signed integer with size 1 byte
'B'	It is a unsigned integer with size 1 byte
'c'	It is a character with size 1 byte
'u'	It is a Unicode character with size 2 bytes
'h'	It is a signed short integer with size 2 bytes
'H'	It is a unsigned short integer with size 2 bytes
'i'	It is a signed integer with size 2 bytes
'I'	It is a unsigned integer with size 2 bytes
'w'	It is a Unicode character with size 4 bytes
'l'	It is a signed integer with size 4 bytes
'L'	It is a unsigned integer with size 4 bytes
'f'	It is a floating-point with size 4 bytes
'd'	It is a floating-point with size 8 bytes

Sample code:

```
from array import *
myArray = array('i,' [1,2,3,4,5])
for i in myArray:
print(i)
```

Output:

```
1
2
3
```

```
4
5
```

Lists as Arrays

As Python do not have a native array data structure, it is required to load the NumPy python module. Both the visual module and the pylab module load NumPy. But, if we use plain python, there is no array. Since arrays look a lot like a list, lists can be employed as arrays. However, arrays (instead of lists) should be used to perform arithmetic operations. Moreover, arrays will store data more compactly and efficiently.

In Python, a one-dimensional array can easily be represented as a list. The following code initializes an array 'myArray' and attempts to find the largest with its items, using the concept with lists in Python.

```
myA=[45, 23, 76, 12, 33]
print("The given elements are")
for i in range(len(myA)):
        print(myA[i])
m=0
for i in range(len(myA)):
        if m<myA[i]:
                m=myA[i]
print("The largest is," m)
```

Output:

The given elements are:

```
45
23
76
12
33
The largest is 76
```

A 2D array can be created using lists within the list. The following code creates the 2×2 matrix as [[11,22],[33,44]] with the list [11,22] representing the first row and the list [33,44] representing the second row.

Sample code:

```
myA=[[11,22],[33,44]]
```

```
for i in range(len(myA)):
        for j in range(len(myA [i])):
                print(myA [i][j])
```

Output:

```
11
22
33
44
```

In a similar manner, a 3×2 matrix with elements ['aa,' 'bb,' 'cc,' 'dd,' 'ee,' 'ff'] is created and displayed along with their indices in the following code:

Sample code:

```
myA=[['aa,"bb'],['cc,"dd'],['ee,"ff']]
for i in range(len(myA)):
        for j in range(len(myA[i])):
                print('[,'i,,"j,'],'myA[i][j])
```

Output:

```
[0 0] aa
[0 1] bb
[1 0] cc
[1 1] dd
[2 0] ee
[2 1] ff
```

7.6. SEARCHING

7.6.1. Sequential Search

Linear search or sequential search is a method for discovering a scrupulous value in a list that scrutiny each element in sequence until the preferred element till the end with the list. It is not compulsory for the list to be ordered.

20	30	40	50	60
0	1	2	3	4

- **Case 1:**
  Search value = 50

- **Step 1:** Compare 50 with value at index 0
- **Step 2:** Compare 50 with value at index 1
- **Step 3:** Compare 50 with value at index 2
- **Step 4:** Compare 50 with value at index 3 (Success)

➢ **Case 2:**

Search value = 70

- **Step 1:** Compare 70 with value at index 0
- **Step 2:** Compare 70 with value at index 1
- **Step 3:** Compare 70 with value at index 2
- **Step 4:** Compare 70 with value at index 3
- **Step 5:** Compare 70 with value at index 4

Failure

Python program for sequential search

```
def seqsearch(ls,item): # define function for sequential search
        pos=0
        found="False"
        while pos<len(ls):
                if(ls[pos]==item):
                        found="True"
                        break
                pos=pos+1
        return(found,pos+1)
ls=[ ]
i=0
n=int(input("Enter number with elements in the list"))
for i in range(0,n,1):
        v=input("Enter the no.")        # Input number
        ls.append(v)
print(ls)
key=input("Enter key to be searched") # Read search key
print(seqsearch(ls,key))        # call function for sequential search
```

Input/Output:

```
Enter number with elements in the list 3
Enter the no. 1
Enter the no. 2
Enter the no. 3
['1,' '2,' '3']
Enter key to be searched 2
```

(True, 2)

7.6.2. Binary Search

A binary search or half-interval search algorithm calculates the target position value within a sorted array. The binary search algorithm is classified as a divide-and-conquer search algorithm and implemented in logarithmic time. The fundamental operation concerned in binary search is as follows.

if (data value == middle)
data value is found
else if (data value < middle)
search the left half with list with the same routine
else
search right half with list with the same routine

- **Case 1:** value == a[mid]
  Value = 11
  low = 0
  high = 8
  mid = (0 + 8) / 2 = 4

1	2	7	9	11	13	17	23	27
0	1	2	3	4	5	6	7	8
↑				↑				↑
low				mid				high

In this case, the key-value we are searching for is positioned in the middle position. The search operation can end there and returns an approximate value (Figure 7.1).

- **Case 2:** val > a[mid]
  val = 23
  low = 0, high = 8
  mid = (0 + 8) / 2 = 4
  new low = mid + 1 = 5

1	2	7	9	11	13	17	23	27
0	1	2	3	4	5	6	7	8
↑				↑	↑			↑
low				mid	n e w low			high

Figure 7.1. Case 2-positon updation.

In this case, the value 23 is larger than the middle. Hence it could be in the right half array that begins from position 5 to position 8. It is shown in Figure 7.1 that the new low is at position 5. The new positions with low and high, are enforced to the determined right half again.

- **Case 3:** val < a[mid]
  val = 7
  low = 0, high = 8
  mid = (0 + 8) / 2 = 4
  new high = mid – 1 =5

1	2	7	9	11	13	17	23	27
0	1	2	3	4	5	6	7	8
↑			↑	↑				↑
low			new high	mid				high

The data value to be searched in this case is 7 that is less than the value at mid position. As the array is sorted, the left half might have the search key. The left half array will begin from the alike starting position low=0 but the high position will be altered to mid – 1, i.e., 3. This algorithm is executed once more on this left half.

Python program for binary search.

```
def binary_search(arr, x):
        low = 0
        high = len(arr) – 1
        mid = 0
        while low <= high:
                mid = (high + low) // 2
                if arr[mid] < x:
                        low = mid + 1
                elif arr[mid] > x:
                        high = mid – 1
                else:
                        return True
                return False
print(binary_search([11,22,33,55,88],32))
function call
print(binary_search([21,22,23,25,28],25))
```

Output:

```
False
True
```

7.6.3. Illustrative Programs

1. Square root using Newtons method:

```
def newtonsqrt(n):
        root=n/2
        for i in range(10):
                root=(root+n/root)/2
        print(root)
n=int(input("Enter number to find Sqrt: "))
newtonsqrt(n)
```

Output:

```
Enter number to find Sqrt: 9
3.0
```

2. GCD with two numbers:

```
x=int(input("Enter the smaller number1:"))
y=int(input("Enter the larger number2:"))
for i in range(1,x+1):
        if(x%i==0 and y%i==0):
                gcd=i
print("GCD is:,"gcd)
```

Output:

```
Enter the smaller number1:8
Enter the larger number2:24
GCD is: 8
```

3. Exponent with number:

```
def powr(base1,exp1):
        if(exp1==1):
                return(base1)
        else:
                return(base1*powr(base1,exp1–1))
base1=int(input("Enter the base number: "))
exp1=int(input("Enter the exponential value:"))
```

```
res=powr(base1,exp1)
print("Result is:,"res)
```

Output:

```
Enter the base number: 2
Enter the exponential value:3
Result is: 8
```

4. Sum with array elements:

```
a=[2,3,4,5,6,7,8]
sum=0
for i in a:
sum=sum+i
print("the sum is,"sum)
```

Output:

```
the sum is 35
```

5. Linear search:

```
a=[20,30,40,50,60,70,89]
print(a)
search=int(input("enter an element to search:"))
for i in range(0,len(a),1):
    if(search==a[i]):
        print("element found at index,"i+1)
        Break
    else:
        print("not found")
```

Output:

```
[20, 30, 40, 50, 60, 70, 89]
enter an element to search:30
element found at index 2
```

6. Binary search:

```
a=[20, 30, 40, 50, 60, 70, 89]
print(a)
search=int(input("enter element to search:"))
```

```
start=0
stop=len(a)–1
while(start<=stop):
        mid=(start+stop)//2
        if(search==a[mid]):
                print("element found at,"mid+1)
                break
        elif(search<a[mid]):
                stop=mid – 1
        else:
                start=mid+1
else:
        print("not found")
```

Output:

```
[20, 30, 40, 50, 60, 70, 89]
enter an element to search:30
element found at 2
```

7. Sample Pascal's triangle:

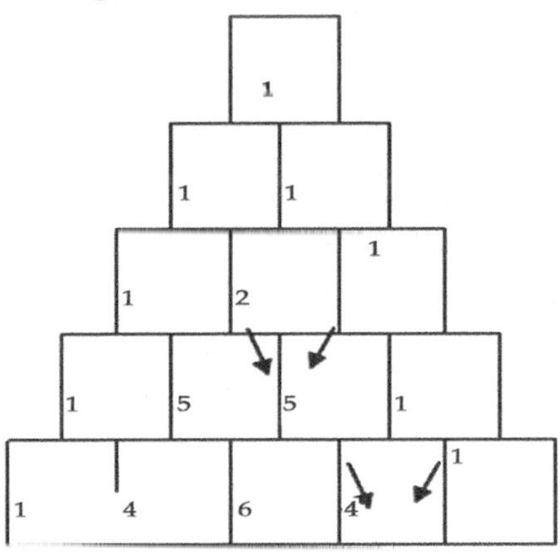

Using recursion:

```
def pascaltri(m):
        if m == 1:
                return [ [1] ]
        else:
                res = pascaltri(m–1)
                last = res[–1]
```

```
            res.append([(x+y) for x,y in zip([0]+last, last+[0])])
            return res
    def display(tree):
        if len(tree) == 0:
            return ''
        else:
            line1 = ' ' * len(tree)
            for cell1 in tree[0]:
                line1 += ' %2i' % cell1
            return line1 + "\n" + display(tree[1:])
    print(display(pascaltri(int(6))))
```

Input/Output:

```
1
1 1
1 2 1
1 3 3 1
1 4 6 4 1
1 5 10 10 5 1
```

Using iteration:

```
    def pascaltri(m):
        row1 = [1]
        a = [0]
        for x in range(max(m,0)):
            print(row1)
            row1=[l+r for l,r in zip(row1+a, a+row1)]
        return m>=1
    pascaltri(6)
```

```
Sample input/output:
[1]
[1, 1]
[1, 2, 1]
[1, 3, 3, 1]
[1, 4, 6, 4, 1]
[1, 5, 10, 10, 5, 1]
```

CHAPTER 8

LISTS

CONTENTS

8.1. LISTS

A list is a sequence of collection of any type of values and be able to be created as a set of comma separated values within square brackets[]. The values in a list are called elements or items. A list inside another list is called nested list.

Sample code for creating lists

```
list1 = ["Ram," "Chennai," 2017]                              # list of different types of elements
list2 = [10, 20, 30, 40, 50]                                  # list of numbers
list3 = []                                                    # empty list
list4 = ["Priya," 2017, 99.8, ["Mumbai," "India"]]            # nested list
print(list1)
print(list2, list3)
print(list4)
```

```
Sample input/output:
["Ram," "Chennai," 2017]
[10, 20, 30, 40, 50] [ ]
["Priya," 2017, 99.8, ["Mumbai," "India"]]
```

8.1.1. Accessing Elements in Lists Using Subscript Operator

The indices of the list's elements are numbered from 0 from the left end and numbered from –1 from the right end. For a list ["Ram," "Chennai," 2017], the indices of elements "Ram," "Chennai" and 2017 are shown in the following table.

Element	"Ram"	"Chennai"	2017
Index from left end	0	1	2
Index from right end	–3	–2	–1

The elements in the list are accessed using the subscript operator [ ] (also known as slicing operator) is used. Index within [ ] indicates the index of the particular element which is present in the list, and it must be an integer expression.

For example, in a list stulist = ["Ram," "Chennai," 2017], stulist[1] returns Chennai as its output.

8.2. LIST OPERATIONS

Lists operate on + and * operators. Here, + represents concatenation and * represents repetition. The following code explains the concatenation of two lists named num1 and num2:

Sample Code and output for + (Concatenation)

Code	Output
num1=[10, 20, 30]	[10, 20, 30]
num2=[40, 50]	[40, 50]
num3=num1+num2	[10, 20, 30, 40, 50]
print(num1)	
print(num2)	
print(num3)	

The following code is another example for concatenation, where a list contains elements of different data types:

Code	Output
stulist = ["Ram," "Chennai," 2017]	["Ram," "Chennai," 2017]
newlist = stulist+["CSE"]	["Ram," "Chennai," 2017, "CSE"]
print(stulist)	
print(newlist)	

Sample code and output for * (Repetition)

The following code describes the repetition operation which is performed on a list num1 for 3 times:

Code	Output
num1=[10, 20]	[10, 20]
num2=num1*3	[10, 20, 10, 20, 10, 20]
print(num1)	
print(num2)	

8.3. LIST SLICES

A part of a list is called a list slice. The operator [m:n] returns the element from the list from mth index to nth index, including the element at mth index but excluding the element at nth index. If the first index is omitted, the elements starting from the beginning will be returned. If the second index is omitted, the elements up to the end will be returned. If the first index value is greater than or equals to the second value, then it returns an empty string. If both indices are omitted, the slice is a given string itself.

Code	Output	Description
stulist = ["Ram," "Chennai," 2017]		List is created with 3 elements
print(stulist[0])	Ram	Slice has the element at index 0
print(stulist[:3])	["Ram," "Chennai," 2017]	Slice is from the beginning
print(stulist[1:])	["Chennai," 2017]	Slice goes to the end
print(stulist[1:1])	[]	Slice is empty
print(stulist[5:2])	[]	Slice is empty
print(stulist[:])	["Ram," "Chennai," 2017]	Entire list is the slice
print(stulist[–2:])	["Chennai," 2017]	Slice goes to the end
print(stulist[:–2])	["Ram"]	Slice is from the beginning
print(stulist[1:3])	["Chennai," 2017]	Slice is from 1st index to 2nd index (excluding the 3rd index)

8.4. LIST METHODS

Python gives the following methods that work on lists:

1. Append: This function is used to add an element to the end of specified list and does not return any value.

Syntax:

```
listname.append(element)
```

Sample code for append:

```
stulist = ["Ram," "Chennai," 2017]
stulist.append("CSE")
print("After appending")
print(stulist)
```

```
Sample output:
["Ram," "Chennai," 2017, "Priya," "Mumbai," 2017]
Count for Chennai: 1
Count for 2017: 2
```

2. Count: This function is used to returns the number of occurrences of an element in a specified list.

Syntax:

```
listname.count(element)
```

Sample code for count:

```
stulist = ["Ram," "Chennai," 2017, "Priya," "Mumbai," 2017]
print(stulist)
print("Count for Chennai:," stulist.count("Chennai"))
print("Count for 2017:," stulist.count(2017))
```

```
Sample output:
["Ram," "Chennai," 2017, "Priya," "Mumbai," 2017]
Count for Chennai: 1
Count for 2017: 2
```

3. Extend: Appends the contents of the second list to the first list and does not return any value.

Syntax:

```
firstlist.extend(secondlist)
```

Sample code for extend:

```
stulist = ["Ram," "Chennai," 2017]
dept = ["CSE"]
print("Before Extend:," stulist)
stulist.extend(dept)
print("After Extend:," stulist)
```

```
Sample output:
Before Extend: ["Ram," "Chennai," 2017]
After Extend: ["Ram," "Chennai," 2017, "CSE"]
```

4. Index: It returns the index of an element, if an element is found in the specified list. Else, an exception is raised.

Syntax:

```
listname.index(element)
```

Sample code for index:

```
stulist = ["Ram," "Chennai," 2017]
print("Index of Ram:," stulist.index("Ram"))
print("Index of Chennai:," stulist.index("Chennai"))
print("Index of 2017:," stulist.index(2017))
```

```
Sample output:
Index of Ram: 0
Index of Chennai: 1
Index of 2017: 2
```

5. Insert: This function is used to inserts the given element at the given index in a specified list and does not return any value

Syntax:

```
listname.insert(index, element)
```

Sample code for insert:

```
stulist = ["Ram," "Chennai," 2017]
print("Before insert:,"stulist)
stulist.insert(1, "CSE")
print("After insert:," stulist)
```

```
Sample output:
Before insert: ["Ram," "Chennai," 2017]
After insert: ["Ram," "CSE," "Chennai," 2017]
```

6. Pop: Removes and returns the element from the end of specified list

Syntax:

```
listname.pop()
```

Sample code for pop:

```
stulist = ["Ram," "Chennai," 2017, "CSE," 92.7]
```

```
print("Initial list is:," stulist)
print("Popping the last item:," stulist.pop())
print("After popping the last item, the list is:," stulist)
```

```
Sample output:
Initial list is: ["Ram," "Chennai," 2017, "CSE," 92.7]
Popping the last item: 92.7
After popping the last item, the list is: ["Ram," "Chennai," 2017, "CSE"]
```

7. Pop(index): Removes and returns the element at given index.

Syntax:

```
listname.pop(index)
```

Sample code for pop(index):

```
stulist = ["Ram," "Chennai," 2017, "CSE," 92.7]
print("Initial list is:," stulist)
print("Popping an item with index 2:," stulist.pop(2))  #2 is item's location to be removed
print("Now the list is:,"stulist)
```

```
Sample output:
Initial list is: ["Ram," "Chennai," 2017, "CSE," 92.7]
Popping an item with index 2: 2017
Now the list is: ["Ram," "Chennai," "CSE," 92.7]
```

8. Remove: Removes an element from the list and does not return any value.

Syntax:

```
listname.remove(element)
```

Sample code for remove:

```
stulist = ["Ram," "Chennai," 2017, "CSE," 92.7, 2017]
print("Initial list is:," stulist)
stulist.remove("CSE")
print("After removing CSE from the list:," stulist)
stulist.remove(2017)
print("After removing 2017 from the list:," stulist)
```

```
Sample output:
Initial list is: ["Ram," "Chennai," 2017, "CSE," 92.7, 2017]
After removing CSE from the list: ["Ram," "Chennai," 2017, 92.7, 2017]
After removing 2017 from the list: ["Ram," "Chennai," 92.7, 2017]
```

9. Reverse: Reverses the entire list

Syntax:

```
listname.reverse()
```

Sample code for reverse:

```
stulist = ["Ram," "Chennai," 2017, "CSE," 92.7]
print("Initial list is:," stulist)
stulist.reverse()
print("After reversing, the list is:," stulist)
```

```
Sample output:
Initial list is: ["Ram," "Chennai," 2017, "CSE," 92.7]
After reversing, the list is: [92.7, "CSE," 2017, "Chennai," "Ram"]
```

10. Sort: Sorts the list in ascending order.

Syntax:

```
listname.sort()
```

Sample code for sort:

Example 1:

```
numlist = [6, 28, 11, 4, 20, 26, 13, 12]
print("Before sorting:," numlist)
numlist.sort()
print("After sorting is:," numlist)
```

Sample output:

```
Before sorting: [6, 28, 11, 4, 20, 26, 13, 12]
After sorting is: [4, 6, 11, 12, 13, 20, 26, 28]
```

Example 2:

```
stulist = ["Ram," "Chennai," CSE,"]
print("Initial list is:," stulist)
```

```
stulist.sort()
print("After sorting, the list is:," stulist)
```

Sample output:

```
Initial list is: ["Ram," "Chennai," "CSE,"]
After sorting, the list is: ["CSE," "Chennai," "Ram"]
```

8.5. LIST LOOP

The elements in the list can be traversed using the "for" loop. The following code shows the exploit of for loop in accessing the elements of a list.

```
numlist = [11, 22, 43, 54, 65]
for j in numlist:
print(j)
```

Output:

```
11
22
43
54
65
```

The following code demonstrates the use of range and len functions in accessing the indices of the elements of a list:

```
numlist1 = [10, 20, 30, 40, 50]
for i in range(len(numlist)):
        print(i)
```

Output:

```
0
1
2
3
4
```

Here, len is a pre-defined function which returns the number of elements present in the list and range is a function that returns a list of index starting from 0 to n – 1, where n is the length (number of elements) of the list. The following code gives an idea to traverse the list and to update the elements of a list with the help of range and len functions in for loop:

```
Numlist1 = [11, 12, 13, 14, 15]
for j in range(len(numlist1)):
        numlist1[j]=numlist1[j]+10
for j in numlist1:
        print(j)
```

Output:

```
21
22
23
24
25
```

A for loop for empty list not at all executes the body and is shown in the following code:

```
numlist = []
for i in numlist:
        print("never executes")
```

8.6. MUTABILITY

The list is a mutable data structure. This means that its elements can be replaced, inserted, and removed. Single or multiple elements of a list can be updated using the slice operator. New elements can be added to the list using the append() method.

The following code replaces "Ram" which is at index 0 in the stulist by "Priya." The values are shown in the output for both instances.

```
stulist = ["Ram," "Chennai," 2017]
print("Before mutation," stulist)
stulist[0] = "Priya"
print("After mutation," stulist)
```

Output:

```
Before mutation ["Ram," "Chennai," 2017]
```

After mutation ["Priya," "Chennai," 2017]

8.7. LIST ALIASING

When we create two lists, we get two objects as exposed in the following code and the corresponding state diagram (Figure 8.1):

```
list1=[11,22]
list2=[31,42]
print(list1 is list2)        # prints False, as list1 and list2 are not the same object
```

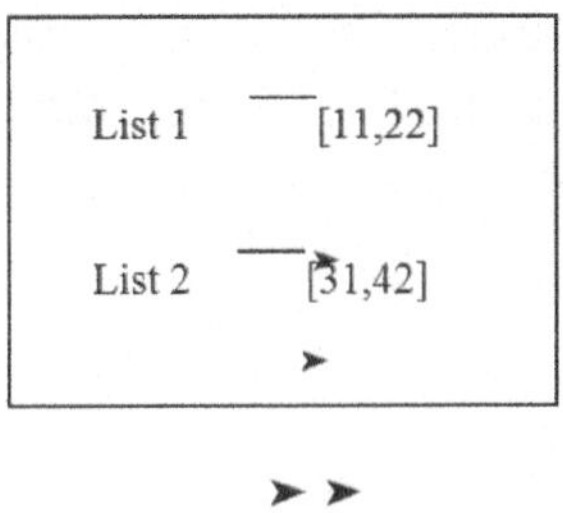

Figure 8.1. State diagram.

Here, the two lists list1 and list2, is correspondent to each other as they contain the same elements, but are not identical because they belong to different objects. If they belong to the same object, then they are identical. But if two objects are the same and equal, then they are not inevitably identical.

In the following code, the same object having two references. We say an object to be aliased when the object with more than one reference has more than one name.

```
first=[11,22]
second=first
print(first is second)        # prints True, as first and
                              second are the same object
```

The state diagram for the above code is as shown in Figure 8.2.

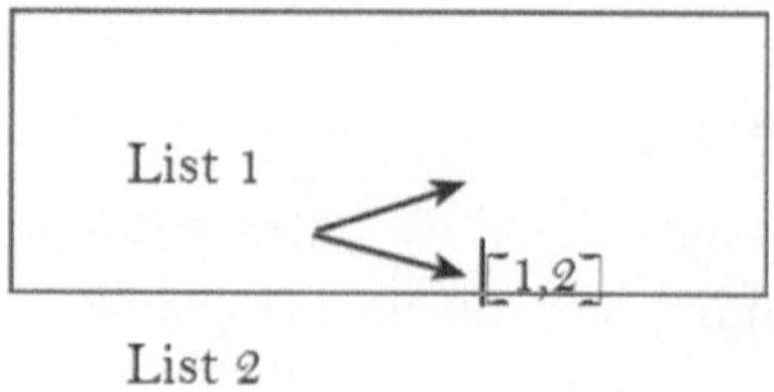

Figure 8.2. State diagram.

If the aliased (copied) object is mutable, modifications done in one object influence the other object also. In the following code, list1 and list2 are aliased objects. Changes made in list1 affect list2 and similarly, changes done in list2 affect list1.

Sample code:

```
first=[11,22]
second=first
print("First is:," first)
print("Second is:," second)
first[0]=10
print("First is:," first)
print("Second is:," second)
second[1]=20
print("First is:," first)
print("Second is:," second)
```

```
Sample output:
First is: [11, 22]
Second is: [11, 22]
First is: [10, 22]
Second is: [10, 22]
First is: [10, 20]
Second is: [10, 20]
```

Though aliasing can be helpful, it may lead to errors. So, avoid aliasing in mutable objects. To prevent aliasing in lists, a new empty list can be created and the contents of the existing list can be copied to it, as given in the following code:

```
list1=[1,2]                          # Existing list
list2=[]                             # New and Empty list
    for e in list1:
```

```
        list2.append(e)
print("List1 is:," list1)
print("List2 is:," list2)
list1[0]=10
print("After modification")
print("List1 is:," list1)
print("List2 is:," list2)
```

```
Output:
        Listl is: [1, 2]
        List2 is: [1, 2]
        After modification
        Listl is: [10, 2]
        List2 is: [1, 2]
```

8.8. CLONING LISTS

Assignment statements in Python is not copying objects. They simply create bindings between two objects. For mutable sequences (like lists), a copy of an existing object may be required, so that one object can be changed without affecting another.

In lists, cloning operation can be used to create a copy of an existing list so that changes made in one copy of the list will not influence another. The copy contains the same elements as the original.

- **Method 1:** list() function:

  Built-in list() function can be used for cloning lists with the following syntax:

  Newlistname = list(Oldlistname)

Sample code:

```
old = [10, 20, 30, 40, 50]
new = list(old)
print("Old list is:," old)
print("New list is:," new)
old[0]=5
print("Old list is:," old)
print("New list is:," new)
```

```
Sample output:
Old list is: [10, 20, 30, 40, 50]
New list is: [10, 20, 30, 40, 50]
Old list is: [5, 20, 30, 40, 50]
New list is: [10, 20, 30, 40, 50]
```

- **Method 2:** copy.copy() function:

Syntax:

Newlistname = copy.copy(Oldlistname)

copy.copy() is little slower than list() since it has to determine data type of old list first.

Sample code:

```
import copy
old = [10, 20, 30, 40, 50]
new = copy.copy(old)  # Returns a shallow copy of old list
print("Old list is:," old)
print("New list is:," new)
old[0]=5
print("Old list is:," old)
print("New list is:," new)
```

```
Sample output:
Old list is: [10, 20, 30, 40, 50]
New list is: [10, 20, 30, 40, 50]
Old list is: [5, 20, 30, 40, 50]
New list is: [10, 20, 30, 40, 50]
```

- **Method 3:** copy.deepcopy() function:

Syntax:

Newname = copy.deepcopy(Oldname)

copy.deepcopy() is the slowest and memory-consuming method.

Sample code:

```
import copy
old = [10, 20, 30, 40, 50]
new = copy.deepcopy(old)      # Returns a deep copy of old
```

```
print("Old list is:," old)
print("New list is:," new)
old[0]=5
print("Old list is:," old)
print("New list is:," new)
```

```
Sample output:
Old list is: [10, 20, 30, 40, 50]
New list is: [10, 20, 30, 40, 50]
Old list is: [5, 20, 30, 40, 50]
New list is: [10, 20, 30, 40, 50]
```

copy() (also known as shallow copy) and deepcopy() differs in the usage of compound objects that are objects containing other objects, like lists). copy() creates a new compound object first and then include references to the objects of the original. deepcopy() constructs a new compound object and then, recursively, inserts copies to the objects of the original. The following code illustrates the use of deepcopy() for a compound (nested) list.

Sample CODE:

```
import copy
old = [1, 2, ["a,""b"]]
new = copy.deepcopy(old)
print("Old list is:," old)
print("New list is:," new)
new[0] = "c"
new[2][1] = "d"
print("Old list is:," old)
print("New list is:," new)
```

```
Sample output:
Old list is: [1, 2, ["a," "b"]]
New list is: [1, 2, ["a," "b"]]
Old list is: [1, 2, ["a," "b"]]
New list is: ["c," 2, ["a," "d"]]
```

8.9. LIST PARAMETERS

When a list is passed as a parameter to a function, the function gets a reference to the list. In the following code, numlist is a list and it is passed as a parameter to my_insert() function. Within my_insert(), it is referenced as t.

```
def my_insert(t):         # function definition
        t.insert(1,15)
numlist = [10, 20, 30, 40, 50]
print("Before calling my_insert function:," numlist)
my_insert(numlist)     # function call
print("After calling my_insert function:," numlist)
```

Here, the parameter t and the variable numlist are denotes the same object. my_insert() function inserts a new element 15 at index 1 in the list. This change is visible to the caller. The elements of a list before and after calling my_insert() are given below as the output:

```
Before calling my_insert function: [10, 20, 30, 40, 50]
After calling my_insert function: [10, 15, 20, 30, 40, 50]
```

The following program employs a function my_display() that creates and returns a new list.

Within my_display(), numlist is referenced as n.

Sample Code:

```
def my_display(n):     # function definition
        return n[:]
numlist = [10, 20, 30, 40, 50]
print("numlist is:," numlist)
newlist=my_display(numlist)  # function call
print("newlist is:," newlist)
```

```
Sample output:
numlist is: [10, 20, 30, 40, 50]
newlist is: [10, 20, 30, 40, 50]
```

The following program includes a function my_display() that creates and displays the elements of a list.

Sample code:

```
def my_display(n):     # function definition
        nlist= n[:]
        print("Within a function:," nlist)
numlist = [10, 20, 30, 40, 50]
print("numlist is:," numlist)
```

```
my_display(numlist)   # function call
```

Sample output:

numlist is: [10, 20, 30, 40, 50]

Within a function: [10, 20, 30, 40, 50]

8.10. DELETING LIST ELEMENTS

The element from the list can be removed using, del operator, if an element to be deleted is known. In the following code, the element "Chennai" is deleted by mentioning its index in the del operator.

```
stulist = ["Ram," "Chennai," 2017, "CSE," 92.7]
print("Initial list is:," stulist)
del stulist[1]
        print("Now the list is:," stulist)
```

Output.

Initial list is: ["Ram," "Chennai," 2017, "CSE," 92.7]

Now the list is: ["Ram," 2017, "CSE," 92.7]

pop() and remove() methods can also be used to delete list elements.

8.11. PYTHON FUNCTIONS FOR LIST OPERATIONS

1. len			
It returns the total number of elements in a list.			
Syntax:	len(listname)		
	Code	Output	
stulist = ["Ram", "Chennai", 2017, "CSE", 92.7]			
print("Length is : ", len(stulist))		Length is : 5	

2. max			
It returns the largest item from the list			
Syntax:	max(listname)		
3. min			
It returns the smallest item from the list			
Syntax:	min(listname)		

8.12. LIST COMPREHENSION

Comprehensions permit sequences to be built from other sequences. It provides a brief way to create lists.

Code	Output
numlist = [6, 28, 11, 4, 20, 26, 13, 12]	
print("Maximum is : ", max(numlist))	Maximum is : 28
print("Minimum is : ", min(numlist))	Minimum is : 4
stulist = ["Anu", "Chennai", "CSE"]	
print("Maximum is : ", max(stulist))	Maximum is : Chennai
print("Minimum is : ", min(stulist))	Minimum is : Anu
4. list	
It Converts a tuple into a list and returns a list.	
Syntax: listname=list(tuplename)	
Code	Output
stu_tuple = ("Anu", "Chennai", 2017, "CSE", 92.7)	Tuple elements : ("Anu", "Chennai",
print("Tuple elements : ", stu_tuple)	2017, "CSE", 92.7)

stulist = list(stu_tuple)	List elements : ["Anu", "Chennai",
print("List elements : ", stulist)	2017, "CSE", 92.7]

A list comprehension contains the following parts:

- Input sequence.
- Variable that is representing members of the input sequence.
- An optional expression.
- An output expression that is producing elements of the output list from members of the Input Sequence that gratify the predicate.

Syntax:

```
[Expression for item in list if condition]
This is equivalent to:
        for item1 in list:
                if conditional:
                        expression
        new = [expression(i) for i in old if filter(i)]
```

New is the resultant list. expression(i) is based on the variable used for each element in the old list. If needed filter can be applied using if statement.

Example:

```
a_list = [1, "4," 9, "a," 0, 4]
squared = [e**2 for e in a_list if type(e) == int]
print(squared)
```

Sample output:
[1, 81, 0, 16]

- The iterator part iterates through each member e of the input sequence a_list.
- The predicate is used to check if the member is an integer.
- If the member is an integer then it is accepted to the output expression, squared, to become a member of the output list.

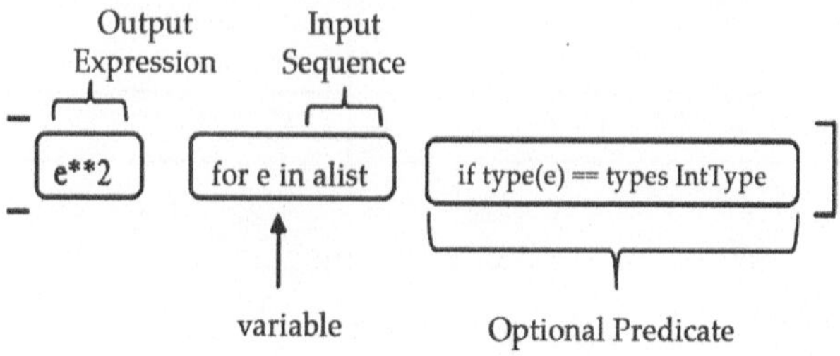

Example:

```
xx=[I for I in range(1,10)]
print(xx)
```

Sample output:

```
[1, 2, 3, 4, 5, 6, 7, 8, 9]
squares=[xx**2 for xx in range(10)]
print(squares)
```

```
Sample output:
[0,1, 4, 9, 16, 25, 36, 49, 64, 81]
```

```
str=["this," "is," "an," "example"]
items=[w[0] for w in str]
print(items)
```

```
Sample output:
["t," "i," "a," "e"]
```

```
value=[xx+yy for xx in [10,20,30] for yy in [1,2,3]]
print(value)
```

```
Sample output:
[11, 12, 13, 21, 22, 23, 31, 32, 33]
```

This example, adds the values in list x to each value in list y.

CHAPTER 9

TUPLES

CONTENTS

9.1. TUPLES

A tuple is a collection of values of different types. Unlike lists, tuple values are indexed by integers values. The difference is that tuples are immutable, i.e., not modifiable.

9.1.1. Advantages of Tuple over List

- Tuples are like lists, so both of them are used in related situations as well. But there are certain advantages of tuples over list, which are scheduled below:
- Tuple generally used for heterogeneous (different) data types, while list for homogeneous (similar) data types.
- Tuples are immutable. Hence iteration done through tuples is quicker when compared to the list.
- Tuple elements could be used as a key element for a dictionary that is not possible with list.

As we have seen before, a tuples are said to be comma-separated values. Syntactically, a tuple can be represented like this:

```
>>> t = “aa,” “bb,” “cc,” “dd,” “ee”
```

Even if it is not necessary to parentheses to enclose tuples, it is so.

```
t1 = (“aa,” “bb,” “cc,” “dd,” “ee”)
t2=(11,22,33,44,55)
```

The empty tuple can be created by simply using the parentheses.

```
t3=()
```

The tuple with one element can be created as below. It takes one comma after the element.

```
t4=(“s,”)
print(type(t4))
```

```
Output:
<type “tuple”>
```

```
t5 = “a,”
print(type(t5))
```

```
Output:
<type “tuple”>
```

A value of tuple in parenthesis is not a tuple. The below program code explains this.

```
t2 = (“aa,”)
print(type(t2[0]))
```

```
Output:
<type “str”>
```

The built-in function tuple can be used to create tuple. To create an empty tuple, no arguments is passed in the built-in function.

```
t = tuple()        # built-in function
print(t)
```

```
Output:
()
```

If the argument is a progression (string, list or tuple), the result is a tuple with the elements of the sequence:

```
tt = tuple(“hello”)
print(tt)
```

```
Output:
(“h,” “e,” “l,” “l,” “o”)
```

```
tt = tuple(“12345”)
print(tt)
```

```
Output:
(“1,” “2,” “3,” “4,” “5”)
```

Program to illustrate the tuple creation for single element

```
#only parentheses is not enough
    tup1 = (“hai”)
```

```
        print(type(tup1))
#need a comma at the end
        tup2 = ("hai,")
        print(type(tup2))
#parentheses is optional
        tup3 = "hai,"
        print(type(tup3))
```

```
Output:
<class "str">
<class "tuple">
<class "tuple">
```

9.1.2. Accessing Values

To access the tuple elements slicing (bracket operator [ ]) operator along with index or indices is used.

```
t1 = ("C," "C++," "python," 1997, 2000);
t2 = (1, 2, 3, 4, 5, 6, 7);
t3= ("a," "b," "c," "d," "e")
print("tup1[0]:," t1[0])
print("tup1[1]:," t1[1])
print("tup2[1:5]:," t2[1:5])
print("tup2[1:]:," t2[1:])
print(t[0])
```

```
Output:
tup1[0]: C
tup1[1]: C++
tup2[1:5]: (2, 3, 4, 5)
tup2[1:]: (2, 3, 4, 5, 6, 7)
a
```

Program to illustrate the accessing the tuple elements

```
t1 = ["p,""y,""t,""h,""o,""n"]
print(t1[0])
print(t1[5])
```

```
print(t1[2]
print t1[-1])
print(t1[-6])
```

nested tuple

```
nesttup = ("hello," [18, 42, 36], (11, 22, 53))
```

nested index

```
print(nesttup[0][4])
print(nesttup[1][2])
print(nesttup[2][0])
```

```
Output:
p
n
t
n
p
o
36
11
```

9.1.3. Updating Tuples

Tuples are immutable means that the tuple values could not be updated or changed. However, the portions of existing tuples are added with a new tuple to create another tuple as the following example demonstrates.

Consider the following tuple:

```
t3= ("aa," "bb," "cc," "dd," "ee")
```

No elements can be modified. If you try to modify, that there occurs an error.

```
t3[1]= "B"
TypeError: "tuple" object does not support item assignment
```

Instead of modifying an element in the tuple sequence, it is obvious to simply replace one tuple with another:

```
t = ("A,") + t3 [1:]
print(t)
```

```
Output:
("A," "bb," "cc," "dd," "ee")
```

Here, the first element "a" is replaced with "A." A new tuple is created with the value "A" is combined with tuple t3 having index from 1 to the last element. The tuple value t3[0]="a" is replaced by "A."

9.1.4. Delete Tuple Elements

It is impossible to delete a single element in the tuple. To delete an entire tuple, the keyword del is used.
Let's consider an example:

```
t = ("C," "C++," "python," 1998, 2001);
print(t)
del(t)
print("After deleting: ")
print(t)
```

```
Output:
("C," "C++," "python," 1998, 2001)
```

After deleting:

```
Traceback (most recent call last):
    File "main.py," line 5, in
    print(t)
    NameError: name "t" is not defined
```

Program for updating and deleting tuples

```
t1 = (2, 3, 4, [5, 6])          #Here [5,6] is a list so it can be changed whereas tuple cannot be changed
print(t1)
```

We cannot change an element

```
t1[3][0] = 7
print(t1)
```

Tuples can be reassigned

```
t1 = ("h,""e,""l,""l,""o")
print(t1)
```

Concatenation

```
print((1, 2, 3) + (4, 5, 6))
```

Repetition operator

```
print(("Repeat,")*3)
delete tuple
del(t1)
print(t1)
```

```
Output:
(2, 3, 4, [5, 6])
(2, 3, 4, [7, 6])
("h," "e," "l," "l," "o")
(1, 2, 3, 4, 5, 6)
("Repeat," "Repeat," "Repeat")
Traceback (most recent call last):
File "<stdin>," line 25, in <module>
print(t1)
NameError: name "t" is not defined
```

9.1.5. Tuple Assignment

Tuple assignment makes it possible to create and assign values for more than one tuple in a single statement itself. For example,

```
x1, y1 = 1, 2
print(x1)
print(y1)
```

```
Output:
1
2
```

In general, to swap the values of two variables, a transitory variable is used. For example, to swap xq and yq:

```
temp = xq
xq = yq
yq = temp
```

This solution is clumsy; the tuple assignment is more elegant.

```
a1, b1 = b1, a1
```

In the expression, the left side belongs to the tuple of variables and the right side belongs to a tuple of expressions. Each value is allocated to its corresponding variable.

The variables on the left and right side of the assignment must be same:

```
a1, b1 = 1, 2
```

In this a1 is assigned with 1 and b1 is assigned with 2.

For the assignment:

```
a1, b1= 1,2,3
```

This statement creates error, as:

```
Value Error: too many values to unpack
```

The right side of the assignment statement can be any kind of sequence (string, list or tuple). For example, to split an email address into a user name and a domain, the split function can be used as follows.

```
mail_id = "students@python.org"
uname, domain = mail_id.split("@")
print(uname)
print(domain)
```

```
Output:
students
python.org
```

In this, the split function is used to separate the value into two parts. The return value from the split function is a list of two elements; the first element is assigned to uname and the second is assigned to domain.

Program to illustrate tuple creation and assignment

```
t1 = () print(t1)
t2 = (11, 21, 13)
print(t2)
t3 = (1, "Hello," 2.4)
print(t3)
t4 = ("World," [18,14,16], (11, 12,1 3))
print(t4)
```

Tuple can be created without parentheses (called tuple packing)

```
t5 = 3, 4.6, "ABC"
print(t5)
```

Tuple unpacking is also possible

```
#tuple assignment
    x, y, z = t5
    print(x)
    print(y)
    print(z)
```

```
Output:
()
(11, 21, 13)
(1, "Hello," 2.4)
("World," [18,1 4,1 6], (11, 12,1 3))
(3, 4.6, "ABC")
3
4.6
ABC
```

9.2. TUPLE METHODS

In python methods for adding items and deleting items are not available. The methods available are count and index.

- count(x) method proceeds the number of occurrences of x
- index(x) method returns index of the first occurrence of the item x.

Program for count and index methods

```
t1 = ("p,""y,""t,""h,""o,""n,""p,""r,""o,""g,""r,""a,""m")
Count print((t1.count("p")))
Index
print(t1.index("y"))
print(t1.index("h"))
```

```
Output:
2
1
3
```

9.3. OTHER TUPLE OPERATIONS

There are some additional tuple operations such as tuple membership test and iterating throughout a tuple. Tuple Membership Test operation can used to check whether an item belongs to a tuple or not. This is done using the keyword in and not in. Iteration through a Tuple operation is performed using a for loop. This could iterate each and every item of a tuple.

Simple program to illustrate tuple operations:

```
t1 = ("p,""y,""t,""h,""o,""n")
```

In operation

```
print("t" in t1)
print("k" in t1)
```

Not in operation

```
print("o" not in t1)
print("b" not in t1)
for lang in ("C,""C++"): print("Progrmming-languages,"lang)
```

```
Output:
True
False
False
True
```

9.4. TUPLES AS RETURN VALUES

In general, a function can return only one value, but if the return value is a tuple, then it is returning multiple values. For example, the procedure for dividing two integers and computing the quotient and remainder could be done by x/y and then x%y. But using python, it could be done at a time. The following code explains this:

```
t = divmod (7, 3)
print(t)
```

```
Output:
(2, 1)
```

Here, the built-in function divmod is used which takes two arguments and returns a tuple of two values, the quotient and remainder. The result can be stored as a tuple as in previous program code. Or tuple assignment can be used to store the elements separately as in the following code.

```
quot, rem = divmod(7, 3)
print(quot)
print(rem)
```

```
Output:
2
1
```

One more example to explain tuples as return values. The built-in functions min and max are used to find the smallest and largest elements of a sequence. The function min_max computes both and returns a tuple of two values.

Here is an example of a function that returns a tuple:

```
def min_max(t):
return min(t), max(t)
```

9.5. BUILT-IN FUNCTIONS WITH TUPLE

Function	Description
all()	Return True if all elements of the tuple are true (or if the tuple is empty).
any()	Return True if any element of the tuple is true. If the tuple is empty, return False.
enumerate()	Assigns an Index to each item in a tuple in python 2 whereas in python3 it just returns memory address value.
len()	Return the length (the number of items) in the tuple.
max()	Return the largest item in the tuple.
min()	Return the smallest item in the tuple
sorted()	Take elements in the tuple and return a new sorted list (does not sort the tuple itself).
sum()	Return the sum of all elements in the tuple.
tuple()	Convert an iterable (list, string, set, dictionary) to a tuple.

9.6. VARIABLE-LENGTH ARGUMENT TUPLES

The functions can take a variable number of arguments for implementation. An argument name that starts with (*) gathers the several arguments into a tuple. For example, printall function takes any number of arguments and prints them:

Example:

```
def printall(*args):                # the function takes several args print (args)
```

The argument name may be anything, but args is conventional. Here is the example to show how the function printall works:

```
n=(1,2.0."3")
printall(n)
(1, 2.0, "3")
```

The complement of gather is scatter. To pass a sequence of values to a function as multiple arguments, the * operator can be used. For example, consider the divmod function which takes exactly two arguments; doesn't work with a tuple of variable length arguments:

```
t = (7, 3)
divmod(t)
```

```
Output:
TypeError: divmod expected 2 arguments, got 1
```

But if you scatter the tuple, it works:

Instead of the above code, the code given below can be used for variable-length arguments.

```
s=divmod(*t)
print (s)
(2, 1)
```

There are some other built-in functions which use variable-length argument tuples.

The max and min functions take any number of arguments:

```
max(1,2,3)
```

```
Output:
3
```

The sum function does not take variable-length arguments. It gives error.

```
sum(1,2,3)
```

```
Output:
TypeError: sum expected at most 2 arguments, got 3
```

9.7. COMPARING TUPLES

With relational operators, it is possible to work with tuples and other sequences. To compare two elements, Python starts by correlating the first element from each sequence. If the elements are equal, it moves to the subsequent elements, and the process continues until it finds an element that is dissimilar. Subsequent elements are not considered (even if they are really big).

```
>>>   (0, 1, 2) < (0, 3, 4)
True
```

The sort function also works in the same way. It sorts primarily by first element. But if there is a tie, it sorts by second element, and so on. This feature lends itself to a pattern called DSU. DSU stands for Decorate, Sort, Undecorate.

9.7.1. DSU

This will Decorate a sequence by constructing a list of tuples with one or more sort keys preceding the elements from the sequence then Sort the list of tuples, and Undecorate by removing the sorted elements of the sequence. For example, to sort a list of words from longest to shortest:

```
def sort_by_length (word1):
        t = []
        for word1 in word1s:
                t.append((len(word1), word1))
                t.sort(reverse=True)
                res1 = []
                for length, word1 in t:
                res1.append(word1)
        return res1
n="Hello World"
x=sort_by_length(n)
print(x)
```

Output:

```
['r,' 'o,' 'o,' 'l,' 'l,' 'l,' 'e,' 'd,' 'W,' 'H,' ' ']
```

The first loop builds a list of tuples, where each tuple is a word preceded by its length. The sort function compares the first element, and its length first, and only considers the second element to break ties. The keyword argument reverse=True tells sort to go in decreasing order. The second loop traverses the list of tuples and builds a list of words in descending order of length.

CHAPTER 10

DICTIONARIES

CONTENTS

10.1. DICTIONARIES

Dictionary is an unordered collection of items. It is similar to a list, but in list elements can be accessed using index, which must be an integer. In Dictionary we access values by looking up a key instead of an index. A key can be any string or number. For example, dictionaries can be used for things like phone books (pairing a name with a phone number), login pages (pairing an e-mail address with a username).

Each item in the dictionary has a key: value pair and the list of items are enclosed inside curly braces {} separated by comma. The values can be of any data type and can repeat; keys must be of immutable types (string, number, or tuple with immutable elements) and must be unique.

Dictionaries in Python are implemented using a hash table. It is an array whose indexes are obtained using a hash function on the keys. A hash function takes a key-value and returns a hash value, an integer. This hash value is used in the dictionary to store and lookup key-value pairs. So keys in dictionary must be hashable.

The following code is a simple example which creates an empty dictionary.

```
my_dict = {}
print(my_dict)
```

```
Sample output:
{}
```

The following dictionary uses integer as keys and string as values.

```
#dictionary with integer keys my_dict = {1: “apple,” 2: “ball”}
print(my_dict)
print(my_dict[2])
```

```
Sample output:
{1: “apple,” 2: “ball”}
ball
```

The following dictionary uses mixed keys. For item 1, both key and its corresponding value are string. In item 2, the key is an integer, and the value is a list.

```
# dictionary with mixed keys
my_dict = {"name": "John," 1: [2, 4, 3]}
print(my_dict)
print(my_dict["name"])
print(my_dict[1])
```

```
Sample output:
{'name': 'John,' 1: [2, 4, 3]}
John
[2, 4, 3]
```

In the output, the order of the key-value pairs is not the same. In general, the order of items in dictionary is unpredictable. In the following example, using list, a mutable data type as key results in the error message.

```
dic = { [1,2,3]:"abc"}
Traceback (most recent call last):
File "main.py," line 1, in <module>
dic = { [1,2,3]:"abc"}
TypeError: unhashable type: "list"
```

Tuple, an immutable data type can be used as a key, which is shown in the following example.

```
my_dic = { (1,2,3):"abc," 3.14:"abc"}
print(my_dic)
```

```
Sample output:
{(1, 2, 3): 'abc,' 3.14: 'abc'}
```

An exception will be raised when we try to access a key that does not exist in the dictionary. In the following example, accessing my_dict[2] results in an error, as the key 2 not exist in the dictionary.

```
my_dict = {"name": "John," 1: [2, 4, 3]}
print(my_dict[2])
```

```
Sample output:
Traceback (most recent call last):
File "main.py," line 2, in <module>
print(my_dict[2])
KeyError: 2
```

Dictionaries can also be created using the dict() function.

```
# using dict()
my_dict = dict({1:"apple," 2:"ball"})
print(my_dict)
```

```
Sample output:
{1: "apple," 2: "ball"}
```

10.2. BUILT-IN DICTIONARY FUNCTIONS AND METHODS

Built-in methods or functions that are available with dictionary are tabulated below.

Function/ Method	Description
len(dict)	Returns the length of the dictionary which is equal to a number of pairs in the dictionary.
sorted(dict)	Returns sorted list of keys in dictionary
dict.clear()	Remove all items from dictionary
dict.copy()	Returns a shallow copy of dictionary
dict. fromkeys(seq[, v])	Return a new dictionary with keys from seq and value equal to v
dict.get(key)	Returns the value of key. If the key does not exists, it returns None

dict.pop(key)	Remove the item with key and returns its value. KeyError occurs when key is not found
dict.popitem()	Remove and return an arbitrary item (key, value). Raises KeyError if the dictionary is empty.
dict.items()	Returns a list of dict"s (key, value) tuple pairs
dict.keys()	Returns list of dictionary dict"s keys
dict1. update(dict2)	Update the dictionary dict1 with the key/ value pairs from dict2, overwriting existing keys.

10.3. ACCESS, UPDATE, AND ADD ELEMENTS IN DICTIONARY

Key can be used either inside square brackets or with the get() method. The difference while using get() is that it returns None instead of KeyError, if the key is not found. Dictionary is mutable.

So we can add new items or change the value of existing items. If the key is present, its value gets updated. Else a new key: value pair is added to dictionary.

```
my_dict={"name":"Ram,""age":21}
print(my_dict) # display all items
print(my_dict.get("name"))# Retrieves value of name key
my_dict["age"]=23      # update value print my_dict
my_dict["dept"]="CSE"          # add item
print(my_dict)
```

```
Sample output:
        {'name': 'Ram,' 'age': 21}
        Ram
        {'name': 'Ram,' 'age': 23, 'dept': 'CSE'}
```

10.4. DELETE OR REMOVE ELEMENTS FROM A DICTIONARY

A picky item in a dictionary can be removed by using the method pop(). This method removes as item with the key provided and returns the value. The method, popitem() can be used to remove and go back an arbitrary item (key, value) from the dictionary. All the items can be detached at once using the clear() method.

```
squares={1:1,2:4,3:9,4:16,5:25}
print(squares.pop(3))  # remove a particular item
print(squares)
print((squares.popitem()))        # remove an arbitrary item print(squares)
del squares[4]  # delete a particular item
print(squares)
squares.clear() # remove all items
print(squares)
```

```
Sample output:
9
{1: 1, 2: 4, 4: 16, 5: 25}
(5, 25)
{1: 1, 2: 4, 4: 16}
{1: 1, 2: 4}
{}
```

We can also use the del keyword to remove individual items or the entire dictionary itself.

If we try to access the deleted dictionary, it will raise an Error.

```
del squares                     # delete the dictionary itself
print(squares)                  #throws error
Traceback (most recent call last):
File "main.py," line 11, in <module>
NameError: name "squares" is not defined
```

10.5. SORTING A DICTIONARY

The items in the dictionary can be sorted using the sorted() function. In the following example, fromkeys() function is used to create a dictionary from a sequence of values. The value 0 is assigned for all keys. Each item is accessed iteratively using for loop that iterate through each key in a dictionary.

```
        mark={}.fromkeys([“Math,””“English,””“Science”],0)
        print(mark)
for item in mark.items():
                print(item)
        print(list(sorted(mark.keys())))
```

```
Sample output:
{“Maths”: 0, “Science”: 0, “English”: 0}
(“Maths,” 0)
(“Science,” 0)
(“English,” 0)
[“English,” “Maths,” “Science”]
```

10.6. ITERATING THROUGH A DICTIONARY

Using a for loop, we can iterate through each key in a dictionary.

```
square={1:1,2:4,3:9,4:16,5:25}
for i in square:
        print(square[i])
```

```
Sample output:
1
4
9
16
25
```

10.7. REVERSE LOOKUP

Lookup is the process of finding the corresponding value for the given key from the dictionary.

It’s easy to find the value given a key to a python dictionary.

```
value=dict[key]
```

Whereas, reverse lookup is the process of finding the key for a given value. There is no direct method to handle reverse lookup. The following function takes a value and returns the first key that maps to that value.

```
def get_Value(dic,value):
        for name1 in dic:
                if dic[name1] == value:
                        return name1
                raise ValueError
squares={1:1,2:4,3:9,4:16,5:25}
print(get_Value(squares,4))                    # successful reverse lookup
```

Sample output:

2

In this example, raise keyword is used to raise/activate an exception. ValueError points that there is a bit wrong with the value of the parameter.

On unsuccessful reverse lookup, when the value is not in the dictionary, the exception ValueError is raised.
Unsuccessful reverse lookup result in following error.

```
print(get_Value(squares,6))                    # unsuccessful reverse lookup

Traceback (most recent call last):
File "main.py," line 7, in <module>
print(get_Value(squares,6))
File "main.py," line 5, in get_Value
raise ValueError
Value Error
```

10.8. INVERTING A DICTIONARY

A dictionary can be inverted with list values. For example, if you were given a dictionary that maps from child to parent, you might want to invert it; that is, create a dictionary that maps from parent to children. Since there might be several children with the same parent, each value in the inverted dictionary should be a list of children.

Example:

```
def invert_dict_nonunique(d):
        newdict = {}
        for k, v in d.items():
                newdict.setdefault(v, []).append(k)
        return newdict
d = {"chi1": "par1,"
        "chi2": "par1,"
        "chi3": "par2,"
        "chi4": "par2,"
        }
print invert_dict_nonunique(d)
```

Sample output:

```
{"par1": ["chi1," "chi2"], "par2": ["chi3," "chi4"]}
```

In this example, the loop iterates through dictionary items where k represents key and v represents value. The setdefault() method will set newdict[v]=[] and append the key value to list..

As we mentioned earlier, the keys in dictionaries have to hashable. It works correctly only when keys are immutable. For example, if key is a list and to store a key-value pair, Python will hash the key and store it in an equivalent location. If that key is modified, it would be hashed to dissimilar location. In this case, we will have two entries for the similar key or might be botched to locate a key. Either way, the dictionary would not work properly. Since lists and dictionaries are mutable, they cannot be used as keys, but they will be used as values.

10.9. MEMOIZATION (MEMOS)

Memoization effectively refers to remembering results of method calling, which is based on the method inputs and then recurring the remembered result pretty than computing the result again.

For example, consider the recursive version to calculate the Fibonacci numbers. The following code to compute the Fibonacci series has an exponential runtime behavior.

Example:

```
def fibo(n):
        if n == 0:
```

```
        return 0
    elif n == 1:
        return 1
    else:
        return fibo(n–1) + fibo(n–2)
```

The runtime behavior of this recursive version can be improved by adding a dictionary to memorize previously calculated values of the function.

```
def memory(f):
    memo = {}
    def helper(x):
        if x not in memo:
            memo[x] = f(x)
        return memo[x]
    return helper
def fib(n):
    if n == 0:
        return 0
    elif n == 1:
        return 1
    else:
        return fib(n–1) + fib(n–2)
fib = memory(fib)
print(fib(6))
# output the 6th number in Fibonacci series (series starts from 0th
position)
```

Sample output:

```
8
```

memory() takes a function as an argument. The function memorize() uses a dictionary "memo" to store the function results. Though the variable "memo" as well as the function "f" are local to memory, they are captured by a closure through the helper function, which is returned as a reference by memory().

So, the call memory(fib) returns a reference to the helper (), which is doing what fib() would do on its own, plus a wrapper which saves the calculated results. For an integer "n" fib(n) will only be called if n is not in the memo dictionary. If it is in it, we can output memo[n] as the result of fib(n).

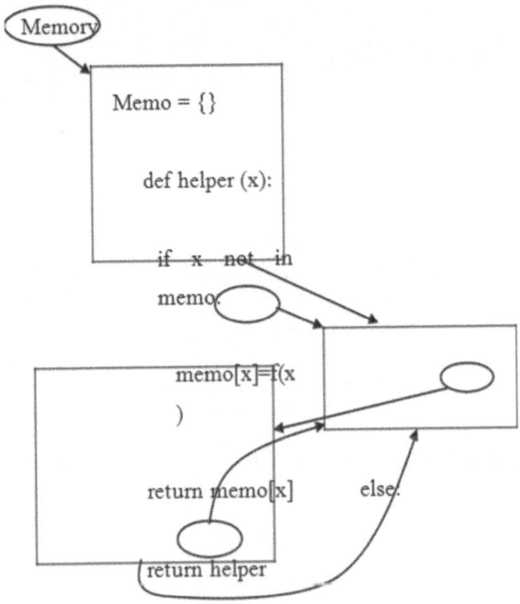

Memo = {}

def helper (x):

if x not in memo:

memo[x]=f(x)

return memo[x]

return helper

fib

```
if n==0:

return 0

elseif n==1:

return 1

else:

return fib(n–1) + fib(n–2)
```

Executing;

```
fib = memory(fib)
```

helper is returned

```
if x not in memo:

memo [x]= f(x)

return memo [x]
```

After executed fib = memory(fib) fib points to the body of the assistant function, which had been returned by memory. The decorated Fibonacci function is called in the return statement return fib(n–1) + fib(n–2), this means the code of the helper function which had been returned by memorize.

10.9.1. Illustrative Programs

1. **Python program to sort a list of elements using the selection sort algorithm:** Selection sort is a sorting algorithm. It is an primed comparison-based algorithm in which the list is divided into two parts: the sorted part at the left end and the unsorted part at the right end. Primarily, the sorted part is empty and the unsorted part is the whole list.

   Selection sort algorithm is comparing first two elements of an array and exchange if compulsory, i.e., if you want to arrange the elements of array in ascending order and if the first element is greater than second at that time, you necessitate to swap the elements but, if the first element is smaller than second, leave the elements as it is. Then, again first element and third element are compared and swapped if compulsory. This process goes on until the first and last element of an array is compared. This completes the first step of selection sort. The working of selection sort algorithm is shown in Figure 10.1.

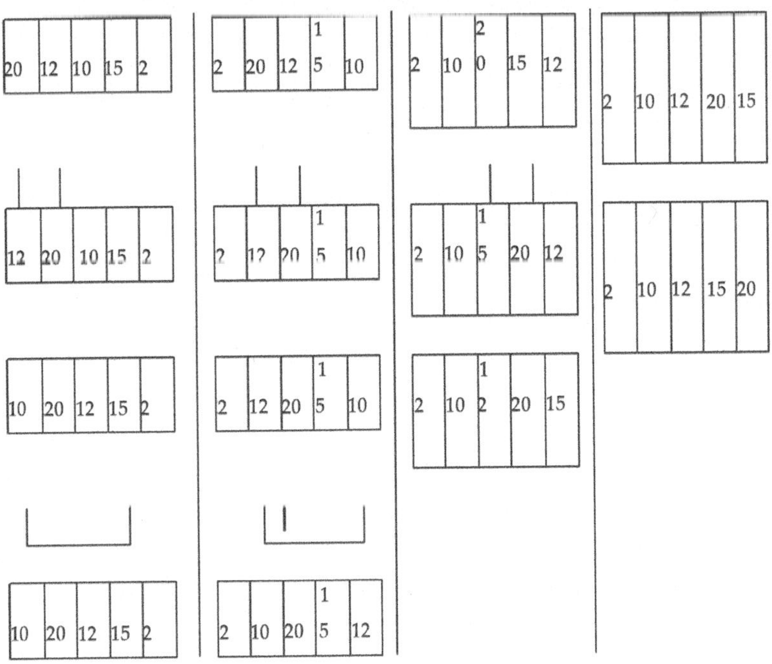

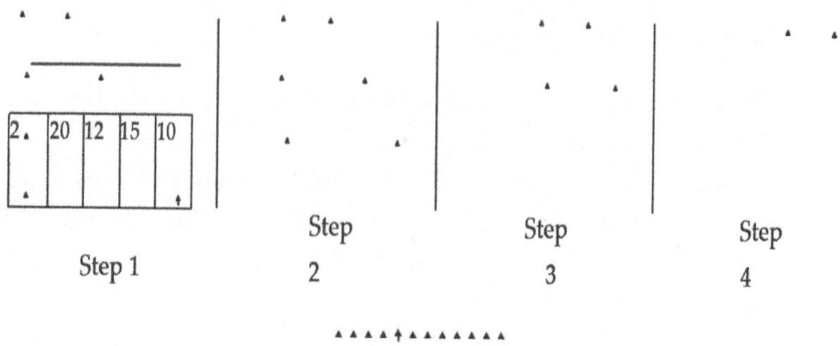

Figure 10.1. Selection sort.

This algorithm is not suitable for large data sets as its average and worst-case complexities are of O(n2), where n is the number of items.

```
list1=[]
n=int(input("Enter number of elements"))
for i in range(0,n):
        x=int(input("Enter number"))
        list1.insert(i,x)
        i+=1
for i in range(len(list1)):
for j in range(i, len(list1)):
        if(list1[i] > list1[j]):
                list1[i], list1[j] = list1[j], list1[i]
print("Sorted List:," list1)
```

```
Sample input/output:
Enter no5
enter no.12
enter no. 2
enter no.23
enter no. 4
enter no.5
Sorted List:
[2, 4, 5, 12, 23]
```

2. **Python program to sort a list of elements using the insertion sort algorithm:** Insertion sort is a simple sorting algorithm which it builds the final sorted array (or list) one item at a time. Here, a

sub-list is at all times sorted. The array is searched consecutively and unsorted items are stimulated and inserted into the sorted sub-list (in the same array). This algorithm is proper for small data sets. But it is much less proficient on large lists than more advanced algorithms such as quicksort, heapsort, or merge sort. The worst case convolution of the algorithm is of O(n2), where n is the number of items (Figure 10.2).

Step 1	12 | 3 | 1 | 5 | 8	Scrutiny second element of array with element Before it and inserting it in appropriate position. In this case, 3 is inserted in position of 12.
Step 2	3 | 12 | 1 | 5 | 8	Scrutiny third element of array with element before it and inserting it in the appropriate position. In this case, 1 is inserted in position of 3.
Step 3	1 | 3 | 12 | 5 | 8	Inspect fourth element of array with element before it and inserting it in the appropriate position. In this case, 5 is inserted in position of 12.

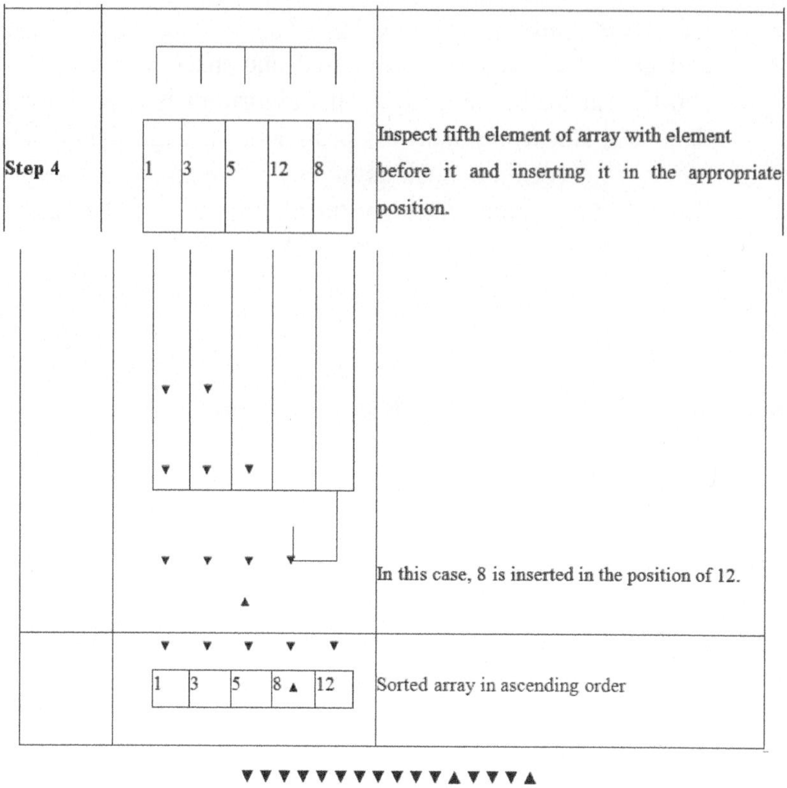

Figure 10.2. Insertion sort.

```
def insertionsort(list1):
        for i in range(1,len(list1)):
                temp=list1[i]
                j=i–1
                while temp<+list1[j] and j>=0:
                        list1[j+1]=list1[j]
                        j=j–1
                        list1[j+1]=temp
        return list1
arr=[]
n=int(input("Enter the number of elements"))
for i in range(0,n):
                        x=int(input("Enter the number"))
                        arr.insert(i,x)
                        i+=1
print(insertionsort(arr))
```

```
Sample input/output:
Enter the number of elements5
Enter the number12
Enter the number23
Enter the number4
Enter the number16
Enter the number34
[4, 12, 16, 23, 34]
```

3. **Python program to sort a list of elements using the merge sort algorithm:** Merge sort is a sorting technique that is similar to the divide and conquer technique. It first divides the array into equal half list and then merges them in a sorted manner. The basic steps caught up in merge sort algorithm are as follows: Given an array A.

   i. **Divide:** If q1 is the half-way point between p1 and r1, then we can split the subarray A[p1..r1] into two arrays A[p1..q1] and A[q1+1, r1].

   ii. **Conquer:** In the conquer step, we try to sort both the subarrays A[p..q] and A[q+1, r]. If we have not yet reached the base case, again divide both these sub arrays and sort them.

   iii. **Combine:** When the surmount step reaches the base step and we get two sorted subarrays A[p1..q1] and A[q1+1, r1] for array A[p1..r1], we merge the results by creating a sorted array A[p1..r1] from two sorted subarrays A[p1..q1] and A[q1+1, r1] (Figure 10.3).

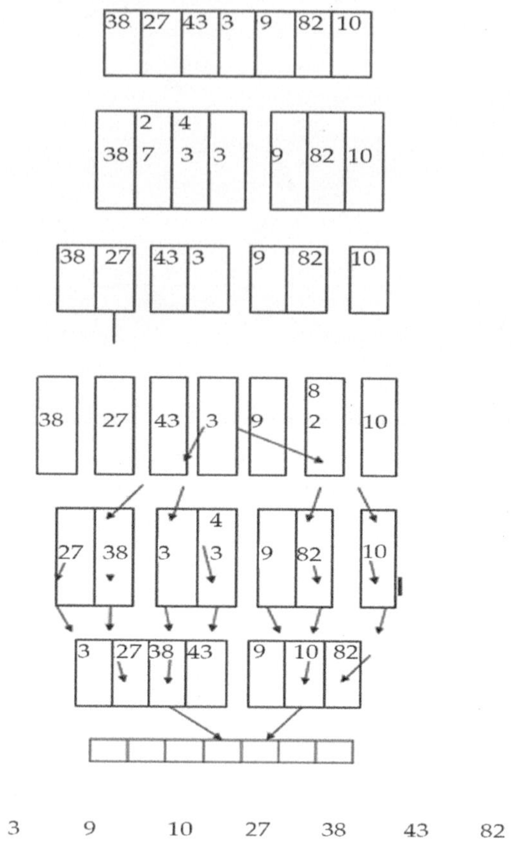

Figure 10.3. Merge sort.

As shown in Figure 10.3, the merge sort algorithm recursively divides the array into halves pending we make the base case of array with 1 element. Later than, the merge function picks up the sorted sub-arrays and merges them to gradually sort the entire array. The worst-case difficulty of the algorithm is O(n log n), where n is the number of items.

```
def merge_sort(seq):
        if len(seq) < 2:
                return seq
        m = int(len(seq) / 2)
        return merge(merge_sort(seq[:m]), merge_sort(seq[m:]))
def merge(lefthalf, lefthalft):
        result = []
```

```
        i = j = 0
        while i < len(lefthalf) and j < len(lefthalft):
                if lefthalf[i] < lefthalft[j]:
                result.append(lefthalf[i])
                i += 1
        else:
                result.append(lefthalft[j])
                j += 1
        result += lefthalf[i:]
        result += lefthalft[j:]
        return result
print(merge_sort([50, 2, 16, 8,15, 80, 11]))
```

```
Sample output:
[2,8,11,15,16, 50, 80]
```

4. **Python program to sort a list of elements using the Quick sort algorithm:** Like Merge Sort, Quick Sort is also a Divide and Conquer algorithm. It picks an element as pivot and partitions the given array around the picked pivot.

The pivot element can be chosen using the following different ways.

- Constantly pick first element as pivot;
- Constantly pick last element as pivot;
- Choose a random element as pivot;
- Choose median as pivot.

The runtime of the algorithm varies based on the pivot selected. The basic idea behind this algorithm is as follows.

- Choose one element in the array as pivot.
- Make one pass throughout the array, called a partition step, re-arranging the entries so that:
  - ➢ The pivot is in its appropriate place;
  - ➢ Entries smaller than the pivot are to the left of the pivot;
  - ➢ Entries larger than the pivot are to its right.
- Recursively apply quick sort to the part of the array that is to the left of the pivot, and to the right part of the array.

The steps involved in quick sort algorithm are listed below and can be understand easily using the example shown in Figure 10.4.

➢ **Step 1:** Prefer the highest index value as pivot.

- **Step 2:** Acquire two variables to point left and right of the list other than pivot.
- **Step 3:** left pointer points to the low index.
- **Step 4:** Right pointer points to the high.
- **Step 5:** While value at left is less than pivot move right.
- **Step 6:** While value at right is greater than pivot move left.
- **Step 7:** If both step 5 and step 6 does not match exchange left and right.
- **Step 8:** If left ≥ right, the point where they met is the new pivot.

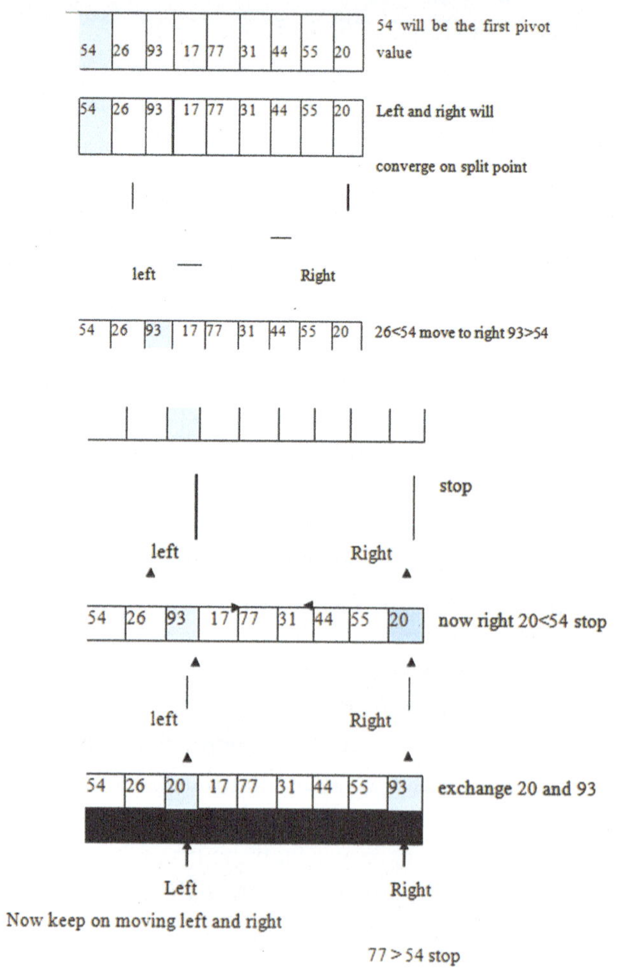

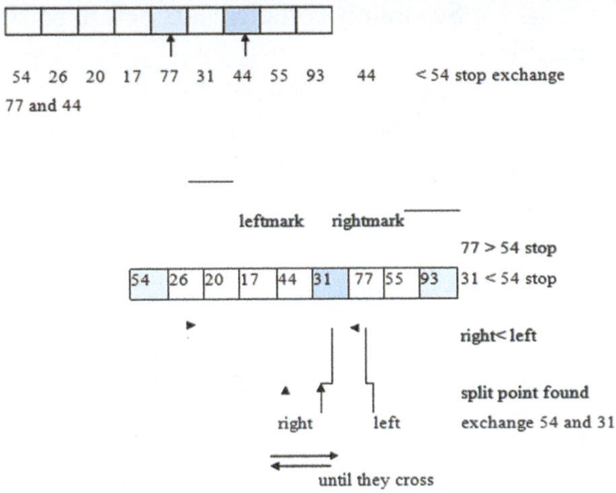

Figure 10.4. Quick sort.

```
        def partition (arr, low, high):
i = (low–1)      # index of smaller element
pivot = arr[high]          # pivot
                for j in range(low, high):
                        # If current element is smaller than or equal to pivot
                        if arr[j] <= pivot:
                                # increment index of smaller element
                                i = i+1
                                arr[i], arr[j] = arr[j], arr[i]
                arr[i+1], arr[high] = arr[high], arr[i+1]
                return (i+1)
        # The main function that implements QuickSort
        # arr[] → Array to be sorted,
        # low → Starting index,
        # high → Ending index
        # Function to do Quick sort
        def quickSort(arr, low, high):
                if len(arr) == 1:
                        return arr
                if low < high:
pi = partition(arr, low, high)    # pi is partitioning index, arr[p] is now at
right place
```

```
            # Separately sort elements before partition and after
        partition
            quickSort(arr, low, pi–1)
            quickSort(arr, pi+1, high)
arr = [10, 7, 8, 9, 1, 5]
n = len(arr)
quickSort(arr, 0, n–1)
print("Sorted array is:," arr)
```

```
Sample output:
Sorted array is: [1, 5, 7, 8, 9, 10]
```

5. Write a Python program to create a histogram from a given list of integers:

```
def histogrm(items):
        for n in items:
            op = ""
            times = n
            while(times > 0):
            op += "*"
            times = times – 1
            print(op)
histogrm([2, 3, 6, 5])
```

```
Sample output:
    **
    ***
    ******
    *****
```

CHAPTER 11

FILES

CONTENTS

11.1. FILES

11.1.1. Persistence

Most of the programs are transient, which means that they run for a short time and produce some output. But, when the program terminates, their data get vanished. When the program started again, it starts with a clean slate. However, there are some other programs which are persistent that they run for a long time (or all the time), maintain at least few of their data in permanent storage (for example, a hard drive) and if the system set to shut down and restart, the program takes the data from where it resides.

For instance, persistent programs are operating systems that run better whenever a computer is on and web servers that run all the time and is waiting for requests to come in on the network.

11.1.2. Reading and Writing Operation

Standard input and output through input functions such as input () and output function print statement are used in file operation.

The Eval Function:

The eval function returns the evaluated result of an input expression.

Input() accepts as string, and eval() evaluates the same.

```
s = input("Enter your input: ")
print(" The output is:," eval(s))
```

```
Output:
Enter your input: [x*5 for x in range(2,10,2)]
Received input is: [10, 20, 30, 40]
```

Now, we will see how to use actual data files. A text file is a sequence of characters stored on a permanent storage medium such as a hard drive, flash memory, or CD-ROM. Python offers some basic functions and methods necessary to manipulate files by default. The basic file operations are open, close, read, and write files.

11.1.2.1. The Open Function

To read or write a file, it is necessary to open it using Python's built-in function named open() function. The open() function creates a file object that could be used to call other methods associated with it. The syntax for the open() function is shown below.

Syntax:

fileobject = open(file_name [, access_mode][, buffering])

The parameters are explained below:

- file_name: The file_name argument is a string value that contains the name of the file to access.
- access_mode: The access_mode denotes the mode in which the file has to be opened (read, write, append, etc.). A complete list of possible values is mentioned in table below. This parameter is optional, and the default file access mode is read (r).
- buffering: If the buffering value is set to 0, then there is no buffering takes place. If the buffering value is 1, then line buffering is performed while accessing a file. If the buffering value is set to an integer greater than 1, then buffering action is performed with the indicated buffer size. If the buffering value is negative, then the buffer size is the system default (default behavior).

The list of different file opening modes:

Modes	Description	
r	Opens a file for reading only. The file pointer is placed at the beginning of the file.	
	This is the default mode.	
rb	Opens a file for reading only in binary format. The file pointer is placed at the beginning of the file. This is the default mode.	
r+	Opens a file for both reading and writing. The file pointer placed at the beginning of the file.	
rb+	Opens a file for both reading and writing in binary format. The file pointer placed at the beginning of the file.	
w	Opens a file for writing only. Overwrites the file if the file exists. If the file does not exist, creates a new file for writing.	
wb	Opens a file for writing only in binary format. Overwrites the file if the file exists. If the file does not exist, creates a new file for writing.	

w+	Opens a file for both writing and reading. Overwrites the existing file if the file exists. If the file does not exist, creates a new file for reading and writing.	
wb+	Opens a file for both writing and reading in binary format. Overwrites the existing file if the file exists. If the file does not exist, creates a new file for reading and writing.	
a	Opens a file for appending. The file pointer is at the end of the file if the file exists.	
	That is, the file is in the append mode. If the file does not exist, it creates a new file for writing.	
ab	Opens a file for appending in binary format. The file pointer is at the end of the file if the file exists. That is, the file is in the append mode. If the file does not exist, it creates a new file for writing.	
a+	Opens a file for both appending and reading. The file pointer is at the end of the file if the file exists. The file opens in the append mode. If the file does not exist, it creates a new file for reading and writing.	
ab+	Opens a file for both appending and reading in binary format. The file pointer is at the end of the file if the file exists. The file opens in the append mode. If the file does not exist, it creates a new file for reading and writing.	

11.1.2.2. The File Object Attributes

Once a file is opened, there would be one file object, to get various information related to that file.

Here is a list of all attributes related to file object:

Attribute	Description
file.closed	Returns true if the file is closed, otherwise false.
file.mode	Returns the file access mode with which file was opened.
file.name	Returns name of the file.

The following illustrate the file attribute description using file object.

```
f = open("file1.txt," "w+")
print("Name of the file:," f.name)
print("Closed or not:," f.closed)
print("Opening mode:," f.mode)
```

```
Output:
Name of the file: file1.txt
Closed or not: False
Opening mode: w+
```

11.1.2.3. The Close() Method

The function close() of a file object flushes if there is any unwritten information and closes the file object when there is no more writing can be done. Python closes a file automatically when the reference object of a file is reassigned with another file. It is a good practice to use the close () method to close a file. The syntax of close () function is given below.

Syntax:

```
fileObject.close();
```

The program to perform the open and close operations of a file.

```
f = open ("file1.txt," "w+")
print("Name of the file:," f.name)
f.close()           #close opened file
print("Closed or not:," f.closed)
```

```
Output:
Name of the file: file1.txt
Closed or not: True
```

11.1.2.4. Reading and Writing Text Files

Python provides read () and write () methods to read and write files through file object, respectively.

11.1.2.5. The Write() Method

The write() method is used to write any string to a file which is opened. Python strings can have binary data and not just text. The write() method does not add a newline character ('\n') to the end of the string. The syntax for write() function is shown below.

Syntax:

```
fileObject.write(string);
```

The argument passed is the content to be written into the opened file. The following program illustrates the file write operation.

```
f= open("file1," "w+")
f.write("Python is a programming language.\nIt is very flexible\n")
f.close()
```

The above program creates a file file1.txt and writes the given content in that file, and finally it closes that file. If the file is opened, then it would have content that is written.

If the file already exists, then opening it in write mode erases the old data and starts fresh. If the file doesn't exist, then a new one is created. The write method is used to put data into the file.
For example:

```
line1 = "Python is a programming language, \n"
f.write (line1)
```

Here, the file object keeps track of where it is pointing to, so if the write function is called again, it adds the new data to the end of the file. For example,

```
line2 = "It is very flexible.\n"
f.write (line2)
```

If no more operations, need to be performed, then the file could be closed using file close() function.

```
f.close()
```

11.1.2.6. The Read() Method

The file read() function reads the file contents from an open file. It is important to note that Python strings can have binary data in addition to text data. The syntax for file read() is given below.

Syntax:

```
fileObject.read([count]);
```

The argument passed is the number of bytes to be read from the opened file. This method starts reading from the beginning of the file, and if the argument count is missing, then it tries to read as much as possible, till the end of the file.

Example to read from the file file1.txten a file:

```
f=open("file1.txt," "w+")
f.write(" Python is a programming language")
f.close()
f = open("file1.txt," "r+")
str = f.read(20);
print(" The string read is:," str)
f.close()
```

Output:

The string read is: Python is a program

11.1.3. Format Operator

The file write() function takes the argument as a string. In order to take other values in a file, it is important to convert them into strings. The easiest way to do is with str function as follows.

```
x = 52
f.write (str(x))
```

Here, the str function converts integer x value as string. An alternative way is to use the format operator, %. When this is applied to integers, % is considered as the modulus operator. But when the first operand is a string, % is considered as the format operator.

The first operand is the format string that contains one or more format sequences, to specify how the second operand is formatted. The result is a string. Consider the simple example.

```
x = 15
print ("%d" % x)
```

```
Output:
15
```

The result is the string '15,' which is not to be confused with the integer value 15. A format sequence can appear anywhere in the string, so you can embed a value in a sentence:

```
bugs= 10
print('I have spotted %d bugs.' % bugs)
```

```
Output:
I have spotted 10 bugs
```

If there is more than one format sequence in the string, the second argument must be a tuple. Each format sequence is matched with an element of the tuple, in sequence. The various format sequences are '%d' to format an integer, '%g' to format a floating-point number and '%s' to format a string.

```
print('In %d years I have spotted %g %s.' % (2, 0.3, 'bugs'))
```

```
Output:
In 2 years I have spotted 0.3 bugs.
```

The number of elements in the tuple has to match the number of format sequences in the string. The types of the elements have to match the format sequences also.

Example:

```
print('%d %d %d' % (1, 2))
```

```
Output:
Traceback (most recent call last):
File "main.py," line 1, in
print '%d %d %d' % (1, 2)
TypeError: not enough arguments for format string
```

Example:

```
print('%d' % 'dollars')
```

```
Output:
Traceback (most recent call last):
File "main.py," line 1, in
print '%d' % 'dollars'
TypeError: %d format: a number is required, not str
```

In the first example, there are three format sequences and only two elements; in the second, the format sequence is for integer, but the element is string.

11.1.3.1. Python File Functions

There are various functions available with the file object.

Method	Description
close()	Close an open file. It has no effect if the file is already closed.
detach()	Separate the underlying binary buffer from the TextIOBase and return it.
fileno()	Return an integer number (file descriptor) of the file.
flush()	Flush the write buffer of the file stream.
isatty()	Return True if the file stream is interactive.
read(n)	Read at most n characters from the file. Reads till end of file if it is negative or None.
readable()	Returns True if the file stream can be read from.
readline(n=–1)	Read and return one line from the file. Reads in at most n bytes if specified.
readlines(n=–1)	Read and return a list of lines from the file. Reads in at most n bytes/characters if specified.
seek(offset,from=SEEK_SET)	Change the file position to offset bytes, in reference to from (start, current, end).

seekable()	Returns True if the file stream supports random access.
tell()	Returns the current file location.
truncate(size=None)	Resize the file stream to size bytes. If size is not specified, resize to the current location.
writable()	Returns True if the file stream can be written to.
write(s)	Write string s to the file and return the number of characters written.
writelines(lines)	Write a list of lines to the file.

11.1.3.2. File Positions

The tell() method provides the current object position within the file. The seek (offset [, from]) method changes the current file position. The offset argument requires the number of bytes to be moved. The from argument states the reference position from where the bytes are to be moved. If from is set to 0, it means use the beginning of the file as the reference position and 1 means use the current position as the reference position and if it is set to 2 then the end of the file would be taken as the reference position. The following program explains the functions of tell() and seek() functions.

```
f=open("file1.txt," "w+")
f.write(" Python is a programming language")
f.close()
f = open("file1.txt," "r+")
str = f.read(20);
print ("The string read is:," str)
pos = f.tell();
print("current file position:," pos)
pos = f.seek(0, 0)
str = f. read (10)
print("Again the string read is:," str)
f. close()
```

```
Output:
The string read is: Python is a program
Current file position: 20
Again the string read is: Python is
```

11.1.3.3. Renaming and Deleting Files

The Python language offers methods that allow us to achieve file-processing operations like renaming and deleting a file. To use this module, it is essential to import the module first, and then the associated functions can be called.

11.1.3.4. The Rename() Method

The rename() method takes two arguments, the current filename and the new filename.

Syntax:

```
os.rename(current_file_name, new_file_name)
```

Example:

Following is the example to rename an existing file file1.txt:

```
import os
Rename a file from file1.txt to file2.txt
os.rename("file1.txt," "file2.txt")
```

11.1.3.5. The Remove() Method

The remove() method can be used to delete files by providing the name of the file to be deleted as the argument.

Syntax:

```
os.remove(file_name)
```

Example:

Following is the example to delete an existing file file2.txt –

```
import os
os.remove("file2.txt")  #Delete file file2.txt
os.mkdir("test")
```

11.1.3.6. Program for Mail Merge

Mail merge is a process of doing this. Instead of writing each mail separately, we have a template for the body of the mail and a list of names that we merge together to form all the mails.

14. Python program to mail merger;
15. Names are in the file names.txt;
16. Body of the mail is in body.txt.

```
#open names.txt for reading with
open("names.txt,""r,"encoding = "utf-8")        #as names_file:

# open body.txt for reading with
open("body.txt,""r,"encoding = "utf-8")         #as body_file:

#read entire content of the body
        body = body_file.read()

#iterate over all names
        for name in names_file:
        mail = "Hello"+name+body

# write the mails to individual files with
open(name.strip()+."txt,""w,"encoding = "utf-8")        #as mail_file:
        mail_file.write(mail)
```

In this program, all the names are written in separate lines in the file named "names.txt." The body of the letter is stored in the file "body.txt"

The two files are opened in reading mode and iterate over each name using a for loop. A new file with the name "[name].txt" is created, where name is the name of that person. Here, the strip() method is used to clean up leading and trailing whitespaces (reading a line from the file also reads the newline '\n' character). Finally, we write the content of the mail into this file using the write() method.

11.1.4. Command Line Arguments

Command line arguments are what we type at the command line prompt along with the script name while we try to execute our scripts. Python-like most other languages provides this feature. sys.argv is a list in Python, which contains the command line arguments passed to the script. We can count the number of arguments using len(sys.argv) function. To use sys.argv, we have to import the sys module.

```
import sys
print ("No. of arguments:," len(sys.argv))
print ("Argument List:,"str(sys.argv))
## this is saved as test.py
```

Run the above script as follows:

```
$ python test.py arg1 arg2 arg3
```

```
Output:
Number of arguments: 4
Argument List: ['test.py,' 'arg1,"arg2,"arg3']
```

11.1.4.1. Filenames and Paths

Files are organized into directories (also called "folders"). Every program has a "current directory," which is the default directory for most operations. The os module provides functions for working with files and directories ("os" stands for "operating system"). os.getcwd () returns the name of the current directory:

```
import os
cwd = os.getcwd()
print(cwd)
```

```
Output:
/web/com/1493114533_4353
```

cwd stands for "current working directory." The result in this example is / web/ com/1493114533_4353.

A string like cwd that identifies a file is called a path. A relative path starts from the current directory; an absolute path starts from the topmost directory in the file system. To obtain the absolute path to a file, you can use os.path.abspath:

```
abs_path=os.path.abspath("file1.txt")
```

```
print(abs_path)
```

```
Output:
/web/com/1493114533_4353/file1.txt
```

os.path.exists checks whether a file or directory exists:

```
print(os.path.exists("file1.txt"))
```

```
Output:
True
```

If it exists, os.path.isdir checks whether it's a directory:

```
print(os.path.isdir("file1.txt"))
```

```
Output:
False
```

Similarly, os.path.isfile checks whether it's a file.

os.listdir returns a list of the files (and other directories) in the given directory:

```
os.listdir(cwd)
['file1,' 'file2']
```

To validate these functions, the following example "walks" through a directory, prints the names of all the files, and calls itself recursively on all the directories.

```
def walk(dirname):
        for name in os.listdir(dirname):
                path = os.path.join(dirname, name)
                if os.path.isfile(path):
                        print path
        else:
                        walk(path)
```

os.path.join takes a directory and a file name and joins them into a complete path.

11.2. ERRORS AND EXCEPTION

11.2.1. Errors

Errors or mistakes in a program are often referred to as bugs. Debugging is the process of finding and eliminating errors. Errors can be classified into three major groups:

1. **Syntax Errors:** Also known as parsing errors are identified by Python while parsing the program. It displays error message and exit without continuing the execution process. Some common Python syntax errors include:
    - Leaving out a keyword;
    - Putting a keyword in the wrong place;
    - Leaving out a symbol, such as a colon, comma or brackets;
    - Misspelling a keyword;
    - Incorrect indentation;
    - Empty block.

    Here are some examples of syntax errors in Python:

    ```
    a=10
    b=20
    if a<b
            print("a is greater")
    ```

    Error Message:

    ```
    File "main.py," line 3
    if a<b
    ^
    ```

    SyntaxError: invalid syntax

    The parser repeats the offending line and displays a little ‘arrow’ pointing at the earliest point in the line where the error was detected. The error is caused by (or at least detected at) the token preceding the arrow: in the example, the error is detected at the if a<b since a colon (‘:’) is missing before it. Filename and line number are printed so you know where to look in case the input came from a script.

```
if True:
        prnt("Hello")
Error Message:
    File "main.py," line 2
    prnt("Hello")
            ^
SyntaxError: invalid syntax
```

In the above example, the error is detected at prnt 'Hello' since print is misspelled.

2. Logical Errors: These occur due to mistake in the program's logic. Here are some examples of mistakes which lead to logical errors:

- Using the wrong variable name;
- Indenting a block to the wrong level;
- Using integer division instead of floating-point division;
- Getting operator precedence wrong;
- Making a mistake in a Boolean expression;
- Off-by-one, and other numerical errors.

Here is an example of logical error in Python:

```
i=1
fact=0
while i<=5:
        fact=fact*i
        i=i+1
print("Fact:," fact)
```

```
Output:
Fact:0
```

In this example for computing factorial of 5, the obtained output is 0. There are no syntax errors. The wrong output occurs due to logical error fact=0. To compute factorial, the fact-value must be initialized to 1. As it is assigned as 0, it results in wrong output.

11.2.2. Exceptions

An exception is an error that occurs during execution of a program. It is also called run time errors. Some examples of Python runtime errors:

- Division by zero
- Performing an operation on incompatible types
- Using an identifier which has not been defined
- Accessing a list element, dictionary value or object attribute which doesn't exist
- Trying to access a file which doesn't exist

An example for run time error is as follows.

```
print (10/0)
Error Message:
Traceback (most recent call last):
File "main.py," line 1, in <module>
print (10/0)
ZeroDivisionError: integer division or modulo by zero
```

Exceptions come in different types, and the type is printed as part of the message: the type in the example is ZeroDivisionError which occurs due to division by 0. The string printed as the exception type is the name of the built-in exception that occurred.

Exception refers to unexpected condition in a program. The unusual conditions could be faults, causing an error which in turn causes the program to fail. The error handling mechanism is referred to as exception handling. Many programming languages like C++, PHP, Java, Python, and many others have built-in support for exception handling.

Python has many built-in exceptions which forces your program to output an error when something in it goes wrong. Some of the standard exceptions available in Python are listed below.

Exception Name	Description
Exception	Base class for all exceptions
ArithmeticError	Base class for all errors that occur for numeric calculation.
OverflowError	Raised when a calculation exceeds the maximum limit for a numeric type.
FloatingPointError	Raised when a floating-point calculation fails.

ZeroDivisionError	Raised when division or modulo by zero takes place for all numeric types.
AssertionError	Raised in case of failure of the Assert statement
EOFError	Raised when there is no input from either the raw_input() or input() function and the end of file is reached.
ImportError	Raised when an import statement fails.
IndexError	Raised when an index is not found in a sequence
KeyError	Raised when the specified key is not found in the dictionary.
NameError	Raised when an identifier is not found in the local or global namespace
IOError	Raised when an input/ output operation fails, such as the print statement or the open() function when trying to open a file that does not exist.
SyntaxError	Raised when there is an error in Python syntax
SystemExit	Raised when Python interpreter is quit by using the sys.exit() function. If not handled in the code, causes the interpreter to exit
TypeError	Raised when an operation or function is attempted that is invalid for the specified data type
ValueError	Raised when the built-in function for a data type has the valid type of arguments, but the arguments have invalid values specified.
RuntimeError	Raised when a generated error does not fall into any category

11.2.3. Handling Exceptions

The simplest way to handle exceptions is with a "try-except" block. Exceptions that are caught in try blocks are handled in except blocks. The exception handling process in Python is shown in Figure 11.1. If an error is encountered, a try block code execution is stopped and control transferred down to except block.

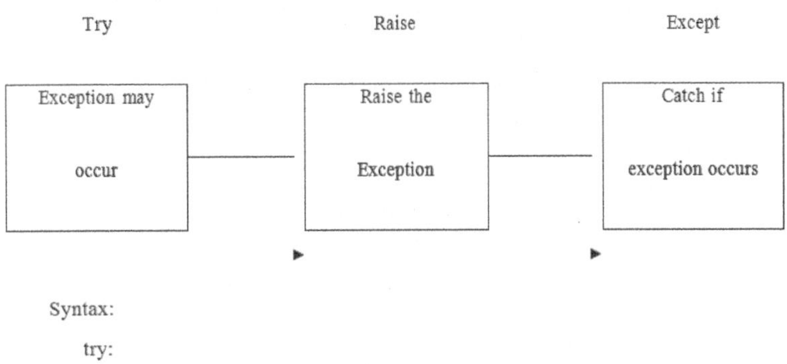

Figure 11.1. Exception handling.

```
1.statements break
        except ErrorName:
        # handler code
```

The try statement works as follows.

- First, the try clause (the statement(s) between the try and except keywords) is executed.
- If no exception occurs, the except clause is skipped and execution of the try statement is finished.
- If an exception occurs during execution of the try clause, the rest of the clause is skipped. Then if its type matches the exception named after the except keyword, the except clause is executed, and then execution continues after the try statement.
- If an exception occurs which does not match the exception named in the except clause, it is passed on to outer try statements; if no handler is found, it is an unhandled exception and execution stops with a message.

A simple example to handle divide by zero error is as follows.

```
(x,y) = (5,0)
try:
z = x/y
```

```
except ZeroDivisionError:
        print "divide by zero"
```

```
Sample output:
divide by zero
```

To display built-in error message of exception, you could have:

```
        (x,y) = (5,0)
        try:
        z = x/y
```

Syntax:

```
        try:
                statements break
                        except ErrorName1:
                                handler code
                        except ErrorName2:
                                handler code
```

A simple example to handle multiple exceptions is as follows.

```
        try:
                dividend = int(input("Please enter the dividend: "))
                divisor = int(input("Please enter the divisor: "))
                print("%d / %d = %f" % (dividend, divisor, dividend/divi-
        sor))
        except ValueError:
                print("The divisor and dividend have to be numbers!")
        except ZeroDivisionError:
                print("The dividend may not be zero!")
```

```
Sample input/output (successful):
Please enter the dividend: 10
Please enter the divisor: 2
10 / 2 = 5.000000
```

```
Sample input/output (unsuccessful-divide by zero error):
Please enter the dividend: 10
Please enter the divisor: 0
The dividend may not be zero!
```

```
Sample input/output (unsuccessful-value error):
Please enter the dividend: 's'
The divisor and dividend have to be numbers!
```

An except clause may name multiple exceptions as a parenthesized tuple, for example:

```
... except (RuntimeError, TypeError, NameError):
...         #handler code
```

Example:

```
try:
        dividend = int(input("Please enter the dividend: "))
        divisor = int(input("Please enter the divisor: "))
        print("%d / %d = %f" % (dividend, divisor, dividend/divi-
sor))
except(ValueError, ZeroDivisionError):
        print("Oops, something went wrong!")
```

```
Sample Input/Output:
Please enter the dividend: 10
Please enter the divisor: 0
Oops,    something    went
```

To catch all types of exceptions using single except clause, simply mention except keyword without specifying error name. It is shown in the following example.

```
try:
        dividend = int(input("Please enter the dividend: "))
        divisor = int(input("Please enter the divisor: "))
        print("%d / %d = %f" % (dividend, divisor, dividend/divi-
sor))
except:
        print("Oops, something went wrong!")
```

11.2.4. Raising Exceptions

The raise statement initiates a new exception. It allows the programmer to force a specified exception to occur. The raise statement does two things: it creates an exception object and immediately leaves the expected program execution sequence to search the enclosing try statements for a matching except clause. It is commonly used for raising user defined exceptions. Two forms of the raise statement are:

Syntax:

```
raise ExceptionClass(value)
raise Exception
```

Example:

```
try:
        raise NameError
except NameError:
        print("Error")
```

```
Sample output:
Error
```

Raise without any arguments is a special use of python syntax. It means get the exception and re-raise it. The process is called re-raise.

Example:

```
try:
        raise NameError('Hi")
except NameError:
        print('Error')
        raise
```

```
Sample output:
Error
Traceback (most recent call last):
File "main.py," line 2, in <module>
raise NameError('Hi')
NameError: Hi
```

In the example, raise statement inside except clause allows you to re-raise the exception NameError.

11.2.5. The Else and Finally Statements

Two clauses that can be added optionally to try-except block are else and finally. else will be executed only if the try clause doesn't raise an exception:

```
try:
        age = int(input("Please enter your age: "))
except ValueError:
        print("Hey, that wasn't a number!")
else:
        print("I see that you are %d years old." % age)
```

```
Sample Input/Output:
Please enter your age: 10
I see that you are 10 years old.
Please enter your age: 'a'
Hey, that wasn't a number!
```

In addition to using except block after the try block, you can also use the finally block. The code in the finally block will be executed regardless of whether an exception occurs and even if we exit the block using break, continue, or return.

```
try:
        age = int(input("Please enter your age: "))
except ValueError:
        print("Hey, that wasn't a number!")
else:
        print("I see that you are %d years old." % age)
finally:
        print("Goodbye!")
```

```
Sample Input/Output:
Please enter your age: 20
I see that you are 20 years old.
Goodbye!
```

11.2.6. User-Defined Exceptions

Python allows the user to create their custom exceptions by creating a new class. This exception class has to be derived, either directly or indirectly, from the Exception class.

Example of Python user-defined exceptions:

```
class Error(Exception):#Base Error
    pass
class PosError(Error):  #Raised when the input value is positive
    pass
class NegError(Error): #Raised when the input value is negative
    pass
number=0
while True:
    try:
        i_num=int(input("Enter a number:"))
        if i_num < number:
            raise NegError
        elif i_num >number:
            raise PosError
            break
    except PosError:
        print("This value is positive!")
    except NegError:
        print("This value is negative!")
```

```
Sample input/output:
Enter a number: 12
This value is positive!
```

In the example, the user defined exception class Error is derived from built-in class Exception. It handles two user defined exceptions: PosError, raised when input value is positive and NegError, raised when input value is negative. The pass keyword indicates null block. The main program reads user input and compares input value with 0. If input>0, the exception PosError is raised using raise keyword else the exception NegError is raised.

CHAPTER 11

MODULES AND PACKAGES

CONTENTS

12.1. MODULES

A Python module is a file that consists of Python code. It allows us to logically arrange related code and makes the code easier to understand and use. It defines functions, classes, and variables.

Python has many useful functions and resources in modules. Functions such as abs() and round() from __builtin__ module are always directly accessible in every Python code. But, the programmer must explicitly import other functions from the modules in which they are defined.

12.1.1. Import Statement

An import statement is used to import Python module in some Python source file.

The syntax is:

```
import module1[, module2[,... modulen]
```

When an 'import' statement is encountered by the interpreter, the corresponding module(s) is imported if it is available in the search path.

Example:

```
import math
```

To use a resource from a module, the following syntax is used:

```
modulename.resourcename
```

For example, math is a built-in module that offers several built-in functions for carrying out basic mathematical operations. The following code imports math module and lists a directory of its resources:

```
import math
print(dir(math))
```

```
Output:
['__doc__,' '__file__,' '__name__,' '__package__,' 'acos,' 'acosh,' 'asin,' 'asinh,' 'atan,' 'atan2,' 'atanh,' 'ceil,' 'copysign,' 'cos,' 'cosh,' 'degrees,' 'e,' 'erf,' 'erfc,' 'exp,' 'expm1,' 'fabs,' 'factorial,' 'floor,' 'fmod,' 'frexp,' 'fsum,' 'gamma,' 'hypot,' 'isinf,' 'isnan,' 'ldexp,' 'lgamma,' 'log,' 'log10,' 'log1p,' 'modf,' 'pi,' 'pow,' 'radians,' 'sin,' 'sinh,' 'sqrt,' 'tan,' 'tanh,' 'trunc']
```

The usage of some built-in functions of math module is shown in the following code along with its output:

```
import math    # Import built-in module math
        print(math.floor(6.9))
        print(math.ceil(6.9))
        print(math.pow(3,4))
```

```
Output:
6.0
7.0
81.0
```

Table 12.1 gives description about a few modules of Python.

Table 12.1. Python Modules and Their Description

Module	Description
cmath	Mathematical operations using complex numbers
copy	Shallow copy and deep copy operations
datetime	Date and time
fileinput	Loop over standard input or list of files
keyword	Testing whether a given string is a keyword
linecache	Accessing individual lines of text files randomly
math	Basic mathematical operations
modulefinder	Finding modules
numbers	Abstract base classes for numerals
operator	Functions analogous to basic operators
py_compile	Compiling Python source code to generate byte code
statistics	Statistical operations
string	String operations

12.1.2. Writing Modules

Any Python source file can be imported as a module into another Python source file. For example, consider the following code named as support.py, which is a Python source file defining two functions add() and display().

```
support.py
def add(a, b):
print("Result is," a+b)
return
def display(p):
print("Welcome,,"p)
return
```

This support.py file can be imported as a module in another Python source file and its functions can be called from the new file as shown in the following code:

```
support.add(4.3)              # calling add() of support module with two integers
support.add(3.5,4.7)          # calling add() of support module with two real values
support.add('a,"b')           # calling add() of support module with two-character value
support.add('Ram,"Kumar')     # calling add() of support module with two string values support.display('Ram')
# calling display() of support module with a string value
```

When this code is executed, the following output is produced:

```
Result is 7
Result is 8.2
Result is ab
Result is RamKumar
Welcome, Ram
from...import Statement
```

It allows us to import specific attributes from a module into the current namespace.

Syntax:

from modulename import name1[, name2[,... nameN]]

The first statement of the following code does not import the entire module support into the current namespace; it just introduces the item add from the module support into the global symbol table of the importing module. Hence, a call to display() function generates an error as shown in the output.

```
from support import add      # Import module support
add(3,4)                     # calling add() of support module with two integer values.
add(3.5,4.7)                 # calling add() of support module with two real values.
add('a,"b')                  # calling add() of support module with two character values.
add('Ram,"Kumar')            # calling add() of support module with two string values.
display('Ram')               # calling display() of support module with a string value
```

```
Output:
Result is 7
Result is 8.2
Result is ab
Result is RamKumar
Traceback (most recent call last):
File "main.py," line 8, in
display('Ram') # calling display() of support module with a string value
NameError: name 'display' is not defined
```

from...import * Statement:

It allows us to import all names from a module into the current namespace.

Syntax:

from modulename import *

Sample code:

```
from support import *  # Import module support
add(3,4)         # calling add() of support module with two integer values
add(3.5,4.7)     # calling add() of support module with two real values.
add('a,"b')      # calling add() of support module with two character values.
add('Ram,"Kumar')   # calling add() of support module with two string values.
display('Ram')# calling display() of support module with a string value.
```

```
Sample output:
Result is 7
Result is 8.2
Result is ab
Result is RamKumar
Welcome, Ram
```

Programs that will be imported as modules often use the following expression:

```
if __name__ == '__main__':
        # test code
```

Here, __name__ is a built-in variable and is set when the program starts execution. If the program runs as a script, __name__ has the value __main__ and the test code is executed. Else, the test code is skipped.

Sample code:

```
        from support import * # Import module support
if __name__ == '__main__':                    # add() and
                                              display()
                                              are called
                                              only if this
                                              pgm runs as
```

script.

```
add(3,4)
display('Ram')
```

Sample output:
Result is 7
Welcome, Ram

reload()

When the module is already imported into a script, the module is not re-read even though it is modified. The code of a module is executed only once. To reload the previously imported module again, the reload() function can be used.

Syntax:

```
reload(modulename)
```

Suppose we have the following code in a module named my_module.

```
print("Welcome")
```

Now, examine the following execution sequence:

- import my_module Welcome
- import my_module
- import my_module

Here, the code is executed only once since the module was imported only once. If the module is subsequently changed, it has to be reloaded. For this reloading, one of the approaches is to restart the interpreter. But this does not help much. So, we can employ reload() inside the imp module as shown:

- import imp
- import my_module Welcome
- import my_module
- imp.reload(my_module) Welcome
  <module 'my_module' from.'\\my_module.py'>

12.1.3. Locating Modules

When a module is imported, Python interpreter searches for the module in the following sequence:

1. The current directory.
2. If the module is not found in the current directory, each directory in the shell variable

   PYTHONPATH is searched.
3. At last, Python checks the installation-dependent default directory.

The module search path is stored in the sys module as sys.path variable. This variable contains the current directory, PYTHONPATH, and the installation-dependent default.

12.2. PACKAGES

When we have a large number of Python modules, they can be organized into packages such that similar modules are placed in one package and different modules are placed in different packages. A package is a hierarchical file directory structure that defines a single Python application environment that consists of modules, sub-packages, sub-subpackages, and so on. In another word, it is a collection of modules. When a package is imported, Python explores in list of directories on sys.path for the package subdirectory.

12.2.1. Steps to Create a Python Package

- Create a directory and name it with a package name.
- Keep subdirectories (subpackages) and modules in it.

  Create __init__.py file in the directory

This __init__.py file can be left empty, but we generally place the initialization code with import statements to import resources from a newly created package. This file is necessary since Python will know that this directory is a Python package directory other than an ordinary directory.

Example:

Assume we are creating a package named Animals with some subpackages as shown in Figure 12.1.

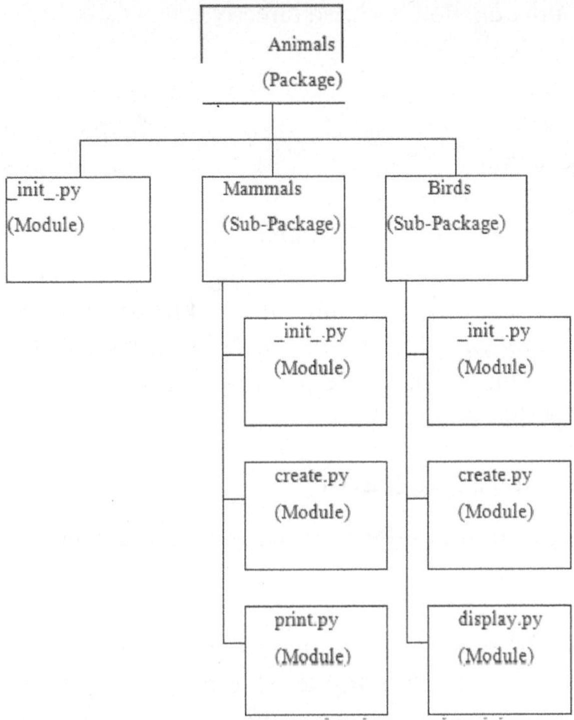

Figure 12.1. Organization of packages and modules.

Modules are imported from packages using dot (.) operator.

- **Method 1:** Consider, we import the display.py module in the above example. It is accomplished by the following statement.

  import Animals.Birds.display

  Now if this display.py module contains a function named display-ByName(), we must use the following statement with full name to reference it.

  Animals.Birds.display.displayByName()

- **Method 2:** On another way, we can import display.py module alone as follows:

  from Animals.Birds import display

Then, we can call displayByName() function simply as shown in the following statement:

  display.displayByName()

- **Method 3:** In the following statement, the required function alone is imported from a module within a package:

  from Animals.Birds.display import displayByName

Now, this function is called directly:

```
displayByName()
```

Though this method is simple, usage of full namespace qualifier avoids confusion and prevents two similar identifier names from colliding.

In the above example, __init__.py of Animals package contains the following code:

```
from Mammals import Mammals
from Birds import Birds
```

In python, module is a single Python file and package is a directory of Python modules containing an additional __init__.py file. Packages can be nested to any depth, but the corresponding directories should include their own __init__.py file.

12.2.2. Illustrative Programs

1. Python program to handle exception when file open fails:

```
try:
        fob = open("test," "r") # open file in read mode
        fob.write("It's my test file to verify exception handling in
Python!!")
except IOError:
print("Error: can\'t find the file or read data") # Exception occurs else:
print("Write operation is performed successfully on the file")
# no Exception
```

Error:

```
        Error: can't find the file or read data
```

2. Python program to raise an exception when the user input is negative

```
try:
        a = int(input("Enter a positive integer value: "))
        if a <= 0:
                raise ValueError("This is not a positive number!!")
        except ValueError as ve:
                print(ve)
```

Sample output:

```
Enter a positive integer value: –1
Error:
```

This is not a positive number!!

3. Python program to count number of words in a file:

```
try:
filename = 'GettysburgAddress.txt'     # specify your input file
        textfile = open(filename, 'w+')
        textfile.write("Hello How are you")
        textfile.close()
        l=0
        with open(filename, 'r') as f:
                for line in f:
                        L=L+len(line.split(" "))
                        print("The number of words are:," L)
                        textfile.close()
except IOError:
        print ("Cannot open file %s for reading" % filename)
        import sys
        sys.exit(0)
```

Sample output:
The number of words are:

4. Python program to count the frequency of words in a text file:

```
file=open("test.txt,""r+")
wordcount={} # define a dictionary that holds words its count
for word in file.read().split():   # for loop iterates through each word is the
file
        if word not in wordcount:
                wordcount[word] = 1
else:
                wordcount[word] += 1
        # Exception is raised
for k,v in wordcount.items():
        print (k,v)
file.close()
```

```
Sample output:
This 1
program 2
example 1
Python 2
```

5. Python program to copy a content of one file to another:

Program 1:

```
rfile=open('testfile.txt,' 'r')              # open file is read mode
try:
        reading_file=rfile.read()
        wfile=open('testfile2.txt,' 'w') # open file is write mode
try:
        wfile.write(reading_file)          # write into file
finally:
        wfile.close()
finally:
        rfile.close()
```

Result:

Contents of testfile will be copied to testfile2 which got created

- **Program 2:**

```
with open("in.txt") as f:
        with open("out.txt," "w") as f1:
                for line in f:
                if "ROW" in line:
                        f1.write(line)
```

Result:

"ROW" will be copied to out.txt which got created.

- **Program 3:** The shutil module offers a number of high-level operations on files and collections of files

```
from shutil import copyfile
copyfile('test.py,' 'abc.py')    # copy content of test.py to
abc.py
```

6. Some Additional Programs: Python program to append text to a file and display the text

```
def file_read(fname):
        with open(fname, "w") as myfile:
                myfile.write("Python Exercises\n")
                myfile.write("Java Exercises")
        txt = open(fname)
        print(txt.read())
file_read('abc.txt')
```

```
Sample output:
Python Exercises
Java Exercises
```

Python program to count the number of lines in a text file

```
def file_lengthy(fname):
        with open(fname) as f:
                for i, l in enumerate(f):
                        pass
        return i + 1
print("Number of lines in the file:,"file_lengthy("test.txt"))
```

Output:
Number of lines in the file: 1

In this example, f is the file object. enumerate(f) iterate over lines of the file. So each time through the loop i gets assigned a line number, and l gets assigned the corresponding line from the file.
Random access file-Python program to read a random line from a file.

A random-access data file enables you to read or write information anywhere in the file. You can use seek() method to set the file marker position and tell() method to get the current position of the file marker. In a sequential-access file, you can only read and write information sequentially, starting from the beginning of the file.

```
f=open('test.txt,"w')
f.write('DearChanna ')
f.seek(4)                    #move file pointer to 4th position from beginning of file
f.write(' Mr.Channa')
f.close()
f=open('Python_source\\test.txt,"r')
s=f.read()
print(s)
```

Sample output:
Dear Mr.Channa

Program that asks the user to input customer information, call writetofile method to write data to the file and call getall method to retrieve customer information from file.

```
          #write data to file
def writetofile(Name,Email=,"Tel=,"Address="):
    try:
          f=open(r'customerlist.txt,"a')
          f.write(Name+':'+Email+':'+Tel+':'+Address)
    except Exception:
          print("error in writing to file...")
    finally:
          f.flush()
          f.close()
#Get all customers'information and display
def getall():
    f=open(r'customerlist.txt,"r')#open file for reading
    content=f.readlines()#read all lines
```

```
        f.close()
                return content
def add():
        Name=input('Name:')
        Email=input('Email:')
        Tel=input('Tel:')
                Address=input('Address:')
                writetofile(Name,Email,Tel,Address)
#main program
add()
print(getall())
```

Output:

```
                Name:xxxxx
                Email:xxx@gmail.com
                Tel:123456789
                Address:India
                ['xxx:xxx@gmail.com:123456789:India']
```

With reference to previous program, write a method to search for a customer by name.

#Search customer by name. If match is found, it returns the position of the name else returns –1.

```
        def search(Name):
global flag     #declare global variable
                try:
                        f=open(r'customerlist.txt,"r')#open file for reading
                        f.seek(0)
                        content=f.readline()
                        while content!=":
                                if content.find(Name)!=–1:
                                        print (content)
                                        flag=1
                                        return                  int(f.tell())-
                                        int(len(content)+1)
                                else:
                                        content=f.readline()
                                        flag=0
```

```
        except Exception:
                print("Error in reading file...")
        finally:
                f.close()
                if flag==0:
print ('Not found')    #Inform the use if the record does not exist
return –1      # The record not found → return –1
    #Main Program
    name=input("enter name to be searched: ")
    search(name)
```

Output:

```
        enter name to be searched: xxxxx
        xxxxx:xxx@gmail.com:123456789:India
```

Using previous search method, program to delete a customer name from file.

```
    def delete(Name):
p=search(Name)      # returns position of customer name.
        print("x=,"p)
if p!=–1:       #Make sure the record exists
st=getall()     #retrieve content from 1st program
f=open(r'customerlist.txt,"w')
f.writelines(st)
f.seek(p)
f.write('*****')        #write 5 starts to override 5 letters of the name to be
else:
print ('No record to delete')   # if the record does not exist
f.close()
name_del=input("enter name to be deleted: ")
delete(name_del)
name_s=input("enter name to be searched: ")
search(name_s)
```

Output:

enter name to be deleted: xxxxx

```
xxxxx:xxx@gmail.com:123456789:India
enter name to be searched: xxxxx
*****:xxx@gmail.com:123456789:India
```

CHAPTER 13

CLASSES IN PYTHON

CONTENTS

13.1. INTRODUCING THE CONCEPT OF CLASSES IN PYTHON

The example for class of parrot can be:

```
class Parrot:
    pass
```

Here, we use the class keyword to define an empty class Parrot. From class, we construct instances. An instance is a specific object created from a particular class.

13.2. OBJECT

An object (instance) is an instantiation of a class. When class is defined, only the description for the object is defined. Therefore, no memory or storage is allocated.

The example for object of parrot class can be:

```
obj = Parrot()
```

Here, obj is an object of class Parrot.

Suppose we have details of parrots. Now, we are going to show how to build the class and objects of parrots.

- *Example 1:* Creating Class and Object in Python

```
class Parrot:
    species = "bird"                    # class attribute
    def __init__(self, name, age):          # instance attribute
        self.name = name
        self.age = age
blu = Parrot("Blu," 10)          # instantiate the Parrot class
woo = Parrot("Woo," 15)
print("Blu is a {}."format(blu.__class__.species))          # access the class attributes
print("Woo is also a {}."format(woo.__class__.species))
print("{} is {} years old."format(blu.name, blu.age))  # access the instance attributes
print("{} is {} years old."format(woo.name, woo.age))
```

Output:

```
Blu is a bird
Woo is also a bird
Blu is 10 years old
```

Woo is 15 years old

In the above program, we created a class with the name Parrot. Then, we define attributes. The attributes are a characteristic of an object.

These attributes are defined inside the __init__ method of the class. It is the initializer method that is first run as soon as the object is created.

Then, we create instances of the Parrot class. Here, blu, and woo are references (value) to our new objects.

We can access the class attribute using __class__. species. Class attributes are the same for all instances of a class. Similarly, we access the instance attributes using blu.name and blu.age. However, instance attributes are different for every instance of a class.

13.3. METHODS

Methods are functions defined inside the body of a class. They are used to define the behaviors of an object.

➢ *Example 2:* Creating Methods in Python

```
class Parrot:
        # instance attributes
        def __init__(self, name, age):
                self.name = name
                self.age = age
                # instance method
        def sing(self, song):
                return "{} sings {}."format(self.name, song)
        def dance(self):
return "{} is now dancing."format(self.name)
                # instantiate the object
blu = Parrot("Blu," 10)
# call our instance methods
print(blu.sing("'Happy'"))
print(blu.dance())
```

Output:

```
Blu sings 'Happy'
Blu is now dancing
```

In the above program, we define two methods, i.e., sing() and dance(). These are called instance methods because they are called on an instance object, i.e., blu.

13.4. INHERITANCE

Inheritance is a way of creating a new class for using details of an existing class without modifying it. The newly formed class is a derived class (or child class). Similarly, the existing class is a base class (or parent class).

- *Example 3:* Use of Inheritance in Python

```
# parent class
class Bird:
        def __init__(self):
        print("Bird is ready")
    def whoisThis(self):
        print("Bird")
    def swim(self):
        print("Swim faster")
child class
class Penguin(Bird):
    def __init__(self):
        # call super() function
        super().__init__()
        print("Penguin is ready")
    def whoisThis(self):
        print("Penguin")
    def run(self):
        print("Run faster")
peggy = Penguin()
peggy.whoisThis()
peggy.swim()
peggy.run()
```

Output:

```
Bird is ready
Penguin is ready
Penguin
Swim faster
Run faster
```

In the above program, we created two classes, i.e., Bird (parent class) and Penguin (child class). The child class inherits the functions of the parent class. We can see this from the swim() method.

Again, the child class modified the behavior of the parent class. We can see this from the whoisThis() method. Furthermore, we extend the functions of the parent class, by creating a new run() method.

Additionally, we use the super() function inside the __init__() method. This allows us to run the __init__() method of the parent class inside the child class.

13.5. ENCAPSULATION

Using OOP in Python, we can restrict access to methods and variables. This prevents data from direct modification which is called encapsulation. In Python, we denote private attributes using underscore as the prefix, i.e., single _ or double __.

➢ *Example 4:* Data Encapsulation in Python

```
class Computer:
        def __init__(self):
                self.__maxprice = 900
        def sell(self):
                print("Selling Price: {}."format(self.__maxprice))
        def setMaxPrice(self, price):
                self.__maxprice = price
c = Computer()
c.sell()
# change the price
c.__maxprice = 1000
c.sell()
        # using setter function
c.setMaxPrice(1000)
c.sell()
```

Output:

```
Selling Price: 900
Selling Price: 900
Selling Price: 1000
```

In the above program, we defined a Computer class.

We used __init__() method to store the maximum selling price of Computer.

We tried to modify the price. However, we can't change it because Python treats the __maxprice as private attributes.

As shown, to change the value, we have to use a setter function, i.e., setMaxPrice() which takes price as a parameter.

13.6. POLYMORPHISM

Polymorphism is an ability (in OOP) to use a common interface for multiple forms (data types).

Suppose, we need to color a shape; there are multiple shape options (rectangle, square, circle). However, we could use the same method to color any shape. This concept is called Polymorphism.

- *Example 5:* Using Polymorphism in Python

```
class Parrot:
        def fly(self):
                print("Parrot can fly")
        def swim(self):
                print("Parrot can't swim")
class Penguin:
        def fly(self):
                print("Penguin can't fly")
        def swim(self):
                print("Penguin can swim")
def flying_test(bird):                          # common interface
        bird.fly()
blu = Parrot()                                  #instantiate objects
peggy = Penguin()
flying_test(blu)                                # passing the object
flying_test(peggy)
```

Output:

```
Parrot can fly
Penguin can't fly
```

In the above program, we defined two classes Parrot and Penguin. Each of them have a common fly() method. However, their functions are different.

To use polymorphism, we created a common interface, i.e., flying_test() function that takes any object and calls the object's fly() method. Thus, when we passed the blu and peggy objects in the flying_test() function, it ran effectively.

➢ *Examples:*

1. Program to convert an integer to a roman numeral:

```
class py_solution:
        def int_to_Roman(self, num):
                val = [
                        1000, 900, 500, 400, 100, 90, 50, 40, 10, 9,
                5, 4, 1]
                syb = ["M," "CM," "D," "CD," "C," "XC," "L,"
        "XL," "X," "IX," "V," "IV," "I"]
                roman_num = "
                i = 0
                while num > 0:
                        for _ in range(num // val[i]):
                                roman_num += syb[i]
                                num– = val[i]
                        i += 1
                return roman_num
print(py_solution().int_to_Roman(15))
print(py_solution().int_to_Roman(2500))
```

Output:

```
XV
MMD
```

2. Program to implement pow(x, n):

```
class py_solution:
        def pow(self, x, n):
                if x==0 or x==1 or n==1:
                        return x
                if x==–1:
                        if n%2 ==0:
                                return 1
                        else:
```

```
                        return –1
            if n==0:
                  return 1
            if n<0:
                  return 1/self.pow(x,-n)
            val = self.pow(x,n//2)
            if n%2 ==0:
                  return val*val
            return val*val*x
print(py_solution().pow(2, –3));
print(py_solution().pow(3, 5));
print(py_solution().pow(100, 0));
```

Output:

```
0.125
243
1
```

3. Program to print output string in uppercase:

```
class IOString():
      def __init__(self):
            self.str1 = ""
      def get_String(self):
            self.str1 = input()
      def print_String(self):
            print(self.str1.upper())
str1 = IOString()
str1.get_String()
str1.print_String()
```

Output:

```
w3resource
W3RESOURCE
```

4. Program to find three elements that sum to zero from a set of n real numbers:

```
class py_solution:
```

```
def threeSum(self, nums):
        nums, result, i = sorted(nums), [], 0
        while i < len(nums) – 2:
                j, k = i + 1, len(nums) – 1
                while j < k:
                        if nums[i] + nums[j] + nums[k] < 0:
                                j += 1
                        elif nums[i] + nums[j] + nums[k] > 0:
                                k– = 1
                        else:
                                result.append([nums[i],     nums[j],
                        nums[k]])
                                j, k = j + 1, k – 1
                                while j < k and nums[j] == nums[j
                        – 1]:
                                        j += 1
                                while j < k and nums[k] == nums[k
                        + 1]:
                                        k  = 1
                i += 1
                while i < len(nums) – 2 and nums[i] == nums[i – 1]:
                        i += 1
        return result
print(py_solution().threeSum([–25, –10, –7, –3, 2, 4, 8, 10]))
```

Output:

```
[[–10, 2, 8], [–7, –3, 10]]
```

5. Program to compute the distance between that point and self:

```
import math
class Point:    #Point class for representing and manipulating x,y
coordinates.
def __init__(self, initX, initY): #Create a new point at the given coordinates.
                self.x = initX
                self.y = initY
        def getX(self):
                return self.x
        def getY(self):
```

```
            return self.y
        def distanceFromOrigin(self):
            return ((self.x ** 2) + (self.y ** 2)) ** 0.5
        def distanceFromPoint(self, otherP):
            dx = (otherP.getX() – self.x)
            dy = (otherP.getY() – self.y)
            return math.sqrt(dy**2 + dx**2)
p = Point(3, 3)
q = Point(6, 7)
print(p.distanceFromPoint(q))
```

Output:

```
5.0
```

6. Program to find the slope of the line joining the origin to the point:

```
class Point:                                          #Point class for representing and manipulating x,y coordinates.
        def __init__(self, initX, initY):             #Create a new point at the given coordinates.
            self.x = initX
            self.y = initY
        def getX(self):
            return self.x
        def getY(self):
            return self.y
        def distanceFromOrigin(self):
            return ((self.x ** 2) + (self.y ** 2)) ** 0.5
        def slope_from_origin(self):
            if self.x == 0:
                return None
            else:
                return self.y / self.x
p = Point(4, 10)
```

```
print(p.slope_from_origin())
```

Output:

```
2.5
```

7. Program to check if a number is palindrome or not:

```
class Check:
        def __init__(self,number):
                self.num = number
        def isPalindrome(self):
                temp = self.num
                result = 0
                while(temp != 0):
                        rem = temp % 10
                result = result * 10 + rem
                temp //= 10
        if self.num == result:
                print(self.num,"is Palindrome")
        else:
                print(self.num,"is not Palindrome")
if __name__ == "__main__":
        check_Palindrome = Check(151)
        check_Palindrome.isPalindrome()
        check_Palindrome = Check(127)
        check_Palindrome.isPalindrome()
```

Output:

```
151 is Palindrome
127 is not Palindrome
```

INDEX